Geometry Supplement

Margaret L. Lial
American River College

Stanley A. Salzman
American River College

Diana L. Hestwood
Minneapolis Community and Technical College

Boston San Francisco New York
London Toronto Sydney Tokyo Singapore Madrid
Mexico City Munich Paris Cape Town Hong Kong Montreal

Reproduced by Addison-Wesley from electronic files supplied by the author.

Copyright © 2002 Pearson Education, Inc.

All rights reserved. No part of this publication may be reproduced, stored in a retrieval system, or transmitted, in any form or by any means, electronic, mechanical, photocopying, recording, or otherwise, without the prior written permission of the publisher. Printed in the United States of America.

ISBN 0-321-09327-5

1 2 3 4 5 6 7 8 9 10   WC   04 03 02 01

Contents

Chapter 8	**Geometry**	**521**
8.1	Basic Geometric Terms	522
8.2	Angles and Their Relationships	529
8.3	Rectangles and Squares	535
8.4	Parallelograms and Trapezoids	545
8.5	Triangles	551
8.6	Circles	559
	Summary Exercises on Perimeter, Circumference, and Area	569
8.7	Volume	571
8.8	Pythagorean Theorem	579
8.9	Similar Triangles	587
	Chapter 8 Summary	595
	Chapter 8 Review Exercises	603
	Chapter 8 Test	611
	Cumulative Review Exercises: Chapters 1–8	613
	Answers	A–27

Preface

This supplement can be used to augment the geometry coverage in any *Basic College Mathematics* textbook, as a brief stand-alone review for those needing basic geometry.

Geometry 8

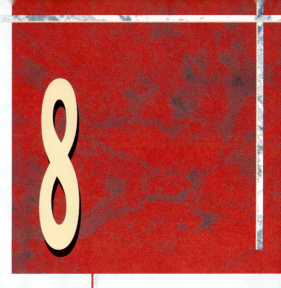

In May 1998, an F-1 tornado (wind speeds of 73 to 112 mph) blew down many large trees in Shoreview, MN. Now, forestry major Bill Masterson is making an inventory of Shoreview's trees to help the city plan replacement plantings. He measures the circumference of each tree at chest height. Then, using the formulas in Section 8.6, he calculates the diameter. This information helps him analyze growth patterns and tree age. (See Exercise 29 in Section 8.6.) (*Source: Shoreview Press.*)

- **8.1** Basic Geometric Terms
- **8.2** Angles and Their Relationships
- **8.3** Rectangles and Squares
- **8.4** Parallelograms and Trapezoids
- **8.5** Triangles
- **8.6** Circles

    Summary Exercises on Perimeter, Circumference, and Area

- **8.7** Volume
- **8.8** Pythagorean Theorem
- **8.9** Similar Triangles

8.1 Basic Geometric Terms

OBJECTIVES

1. Identify lines, line segments, and rays.
2. Identify parallel and intersecting lines.
3. Identify and name angles.
4. Classify angles as right, acute, straight, or obtuse.
5. Identify perpendicular lines.

Geometry developed centuries ago when people needed a way to measure land. The name *geometry* comes from the Greek words *ge*, meaning earth, and *metron*, meaning measure. Today we still use geometry to measure farmland. It is also important in architecture, construction, navigation, art and design, physics, chemistry, and astronomy. You can use it at home when you buy carpet or wallpaper, hang a picture, or build a fence. This chapter discusses the basic terms of geometry and the common geometric shapes that are all around us.

Geometry starts with the idea of a point. A **point** is a location in space. It has no length or width. A point is represented by a dot and is named by writing a capital letter next to the dot.

Point *P*

1 **Identify lines, line segments, and rays.** A **line** is a straight row of points that goes on forever in both directions. A line is drawn by using arrowheads to show that it never ends. The line is named by using the letters of any two points on the line.

Line *AB*, written $\overleftrightarrow{AB}$

A piece of a line that has two endpoints is called a **line segment**. A line segment is named using its endpoints. The segment with endpoints *P* and *Q* is shown below. It can be named $\overline{PQ}$ or $\overline{QP}$.

Line segment *PQ*, written $\overline{PQ}$

A **ray** is a part of a line that has only one endpoint and goes on forever in one direction. A ray is named by using the endpoint and some other point on the ray. The endpoint is always mentioned first.

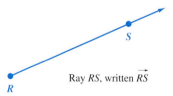
Ray *RS*, written $\overrightarrow{RS}$

Example 1 Identifying Lines, Rays, and Line Segments

Identify each figure as a line, line segment, or ray.

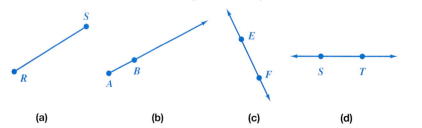

(a) (b) (c) (d)

Figure **(a)** has two endpoints, so it is a line segment.
Figure **(b)** starts at point *A* and goes on forever in one direction, so it is a ray.
Figures **(c)** and **(d)** go on forever in both directions, so they are lines.

Work Problem ➊ **at the Side.**

➊ Identify each figure as a line, line segment, or ray.

(a)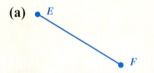

(b)

(c)

(d)

ANSWERS

1. (a) line segment (b) ray (c) line
   (d) line segment

2 Identify parallel and intersecting lines.

A *plane* is an infinitely large flat surface. A floor or a wall is a part of a plane. Lines that are in the same plane, but that never intersect (never cross), are called **parallel lines,** while lines that cross are called **intersecting lines.** (Think of an intersection, where two streets cross each other.)

Example 2 Identifying Parallel and Intersecting Lines

Label each pair of lines as parallel or intersecting.

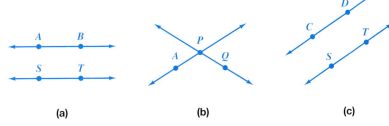

The lines in Figures **(a)** and **(c)** never intersect. They are parallel lines. The lines in Figure **(b)** cross at *P*, so they are intersecting lines.

> **Work Problem ❷ at the Side.**

3 Identify and name angles.

An **angle** is made up of two rays that start at a common endpoint. This common endpoint is called the *vertex*.

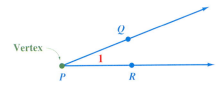

The rays *PQ* and *PR* are called *sides*. The angle can be named in four different ways, as shown below.

Naming an Angle

When naming an angle, the vertex is written alone or it is written in the middle of two other points. If two or more angles have the **same vertex,** as in Example 3 below, do **not** use the vertex alone to name an angle.

Example 3 Identifying and Naming an Angle

Name the highlighted angle.

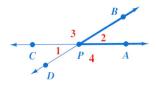

The angle can be named ∠*BPA*, ∠*APB*, or ∠2. It cannot be named ∠*P*, using the vertex alone, because four different angles have *P* as their vertex.

> **Work Problem ❸ at the Side.**

❷ Label each pair of lines as parallel or intersecting.

(a)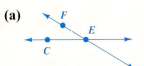

(b)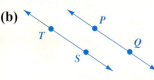

(c)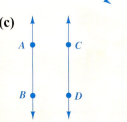

❸ (a) Name the highlighted angle in three different ways.

(b) Darken the rays that make up ∠*ZTW*.

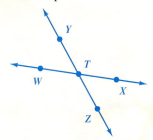

(c) Name this angle in four different ways.

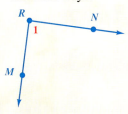

ANSWERS
2. (a) intersecting (b) parallel (c) parallel
3. (a) ∠3, ∠*CQD*, ∠*DQC*
   (b)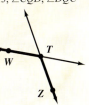

   (c) ∠1, ∠*R*, ∠*MRN*, ∠*NRM*

4 **Classify angles as right, acute, straight, or obtuse.** Angles can be measured in **degrees**. The symbol for degrees is a small, raised circle °. Think of the minute hand on a clock as a ray of an angle. Suppose it is at 12:00. During one hour of time, the minute hand moves around in a complete circle. It moves 360 *degrees*, or 360°. In half an hour, at 12:30, the minute hand has moved half way around the circle, or 180°. An angle of 180° is called a **straight angle.** When two rays go in opposite directions, the rays form a straight angle.

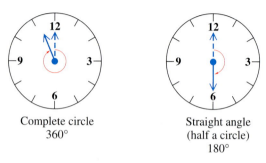

In a quarter of an hour, at 12:15, the minute hand has moved $\frac{1}{4}$ of the way around the circle, or 90°. An angle of 90° is called a **right angle.** Sometimes you hear it called a *square angle.* The minute hands at 12:00 and 12:15 form one corner of a square. So, to show that an angle is a **right angle**, we draw a **small square** at the vertex.

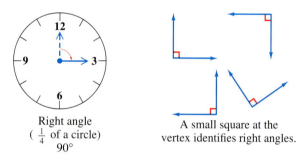

In one minute, the minute hand moves 6°. From this you can tell that an angle of 1° is very small.

Some other terms used to describe angles are shown below.

Acute angles measure less than 90°.

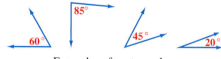

Examples of acute angles

Obtuse angles measure more than 90° but less than 180°.

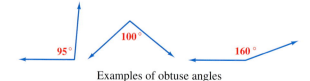

Examples of obtuse angles

Section 10.1 shows you how to use a tool called a *protractor* to measure the number of degrees in an angle.

Classifying Angles

Acute angles measure less than 90°.
Right angles measure exactly 90°.
Obtuse angles measure more than 90° but less than 180°.
Straight angles measure exactly 180°.

NOTE

Angles can also be measured in radians, which you will learn about in a later math course.

Example 4 Classifying Angles

Label each angle as acute, right, obtuse, or straight.

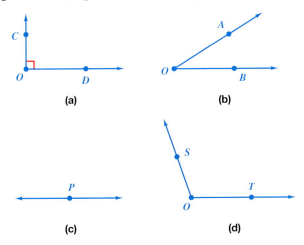

Figure **(a)** shows a *right angle* (exactly 90° and identified by a small square at the vertex).
Figure **(b)** shows an *acute angle* (less than 90°).
Figure **(c)** shows a *straight angle* (exactly 180°).
Figure **(d)** shows an *obtuse angle* (more than 90° but less than 180°).

= **Work Problem ❹ at the Side.**

5 **Identify perpendicular lines.** Two lines are called **perpendicular lines** if they intersect to form a right angle.

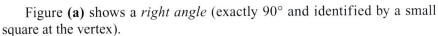

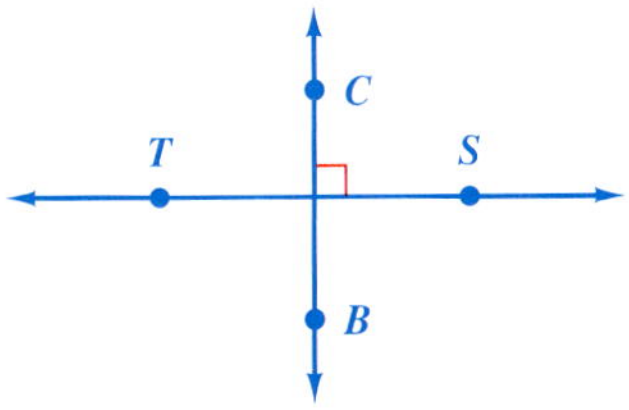

Lines *CB* and *ST* are **perpendicular** because they intersect at right angles. This can be written in the following way: $\overleftrightarrow{CB} \perp \overleftrightarrow{ST}$.

❹ Label each figure as an acute, right, obtuse, or straight angle.

(a)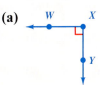

(b)

(c)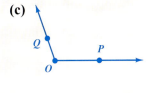

(d)

Answers
4. (a) right (b) straight (c) obtuse
   (d) acute

5 Which pair of lines is perpendicular? How can you describe the other pair of lines?

(a)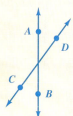

(b)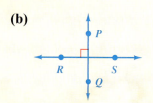

Example 5 Identifying Perpendicular Lines

Which pairs of lines are perpendicular?

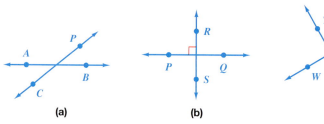

(a) (b) (c)

The lines in Figures **(b)** and **(c)** are perpendicular to each other because they intersect at right angles.

The lines in Figure **(a)** are intersecting lines, but they are not perpendicular because they do not form a right angle.

Work Problem 5 at the Side.

ANSWERS

5. Figure **(b)** shows perpendicular lines; Figure **(a)** shows intersecting lines.

8.1 Exercises

Name each line, line segment, or ray using the appropriate symbol. See Example 1.

1.
2.
3.
4.
5.
6.

Label each pair of lines as parallel, perpendicular, or intersecting. See Examples 2 and 5.

7.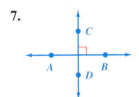
8.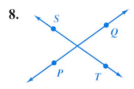
9.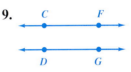
10.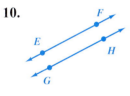
11.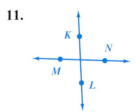
12.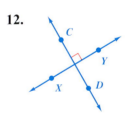

Name each highlighted angle by using the three-letter form of identification. See Example 3.

13.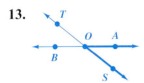
14.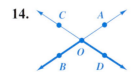
15.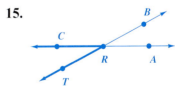
16.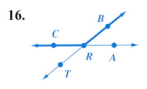
17.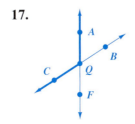
18.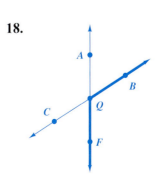

Label each angle as acute, right, obtuse, or straight. For right and straight angles, indicate the number of degrees in the angle. See Example 4.

19.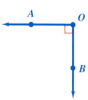

20.

21.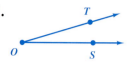

22.

23.

24.

25. Explain what is happening in each sentence.
    (a) The road was so slippery that my car did a 360.
    (b) After the election, the governor's view on taxes took a 180° turn.

26. Find at least four examples of right angles in your home, at work, or on the street. Make a sketch of each example and label the right angle.

RELATING CONCEPTS (Exercises 27–32) FOR INDIVIDUAL OR GROUP WORK

Use the figure below to **work Exercises 27–32 in order.** *Decide whether each statement is* **true** *or* **false.** *If it is true, explain why. If it is false, rewrite to make it a true statement.*

27. ∠UST is 90°.

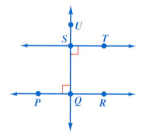

28. $\overleftrightarrow{SQ}$ and $\overleftrightarrow{PQ}$ are perpendicular.

29. The measure of ∠USQ is less than the measure of ∠PQR.

30. $\overleftrightarrow{ST}$ and $\overleftrightarrow{PR}$ are intersecting.

31. $\overleftrightarrow{QU}$ and $\overleftrightarrow{TS}$ are parallel.

32. ∠UST and ∠UQR measure the same number of degrees.

8.2 Angles and Their Relationships

1 **Identify complementary angles and supplementary angles.** Two angles are called **complementary angles** if their sum is 90°. If two angles are complementary, each angle is the *complement* of the other.

OBJECTIVES

1 Identify complementary angles and supplementary angles.

2 Identify congruent angles and vertical angles.

Example 1 Identifying Complementary Angles

Identify each pair of complementary angles.

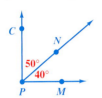

∠MPN (40°) and ∠NPC (50°) are complementary angles because

$$40° + 50° = 90°.$$

∠CAB (30°) and ∠FHG (60°) are complementary angles because

$$30° + 60° = 90°.$$

— Work Problem **1** at the Side.

1 Identify each pair of complementary angles.

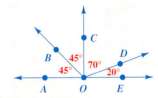

Example 2 Finding the Complement of Angles

Find the complement of each angle.

(a) 30°
The complement of 30° is 60°, because **90°** − 30° = 60°.

(b) 40°
The complement of 40° is 50°, because **90°** − 40° = 50°.

— Work Problem **2** at the Side.

2 Find the complement of each angle.

(a) 35°

(b) 80°

Two angles are called **supplementary angles** if their sum is 180°. If two angles are supplementary, each angle is the *supplement* of the other.

Example 3 Identifying Supplementary Angles

Identify each pair of supplementary angles.

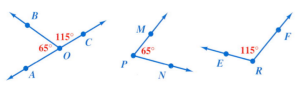

∠BOA and ∠BOC, because 65° + 115° = **180°**.
∠BOA and ∠ERF, because 65° + 115° = **180°**.
∠BOC and ∠MPN, because 115° + 65° = **180°**.
∠MPN and ∠ERF, because 65° + 115° = **180°**.

— Work Problem **3** at the Side.

3 Identify each pair of supplementary angles.

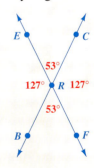

ANSWERS
1. ∠AOB and ∠BOC; ∠COD and ∠DOE
2. (a) 55° (b) 10°
3. ∠CRF and ∠BRF; ∠CRE and ∠ERB; ∠BRF and ∠BRE; ∠CRE and ∠CRF

4 Find the supplement of each angle.

(a) 175°

(b) 30°

Example 4 Finding the Supplement of Angles

Find the supplement of each angle.

(a) 70°
The supplement of 70° is 110°, because **180°** − 70° = 110°.

(b) 140°
The supplement of 140° is 40°, because **180°** − 140° = 40°.

Work Problem **4** at the Side.

2 **Identify congruent angles and vertical angles.** Two angles are called **congruent angles** if they measure the same number of degrees. If two angles are congruent, this is written as ∠A ≅ ∠B and read as, "angle A **is congruent to** angle B." Here is an example.

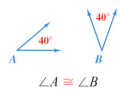

∠A ≅ ∠B

Example 5 Identifying Congruent Angles

Identify the angles that are congruent.

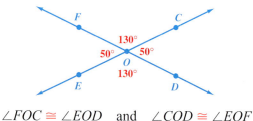

∠FOC ≅ ∠EOD and ∠COD ≅ ∠EOF

Work Problem **5** at the Side.

5 Identify the angles that are congruent.

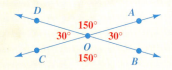

Angles that do not share a common side are called *nonadjacent* angles. Two nonadjacent angles formed by intersecting lines are called **vertical angles**.

Example 6 Identifying Vertical Angles

Identify the vertical angles in this figure.

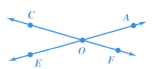

6 Identify the vertical angles.

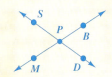

∠AOF and ∠COE are vertical angles because they do not share a common side and they are formed by two intersecting lines ($\overleftrightarrow{CF}$ and $\overleftrightarrow{EA}$).

∠COA and ∠EOF are also vertical angles.

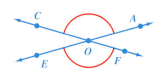

Work Problem **6** at the Side.

ANSWERS
4. (a) 5° (b) 150°
5. ∠BOC ≅ ∠AOD; ∠AOB ≅ ∠DOC
6. ∠SPB and ∠MPD; ∠BPD and ∠SPM

Look back at Example 5 on the previous page. Notice that the two *congruent* angles that measure 130° are also *vertical* angles. Also, the two congruent angles that measure 50° are vertical angles. This illustrates the following property.

Congruent Angles

If two angles are vertical angles, they are **congruent**, that is, they measure the same number of degrees.

Example 7 Finding the Measures of Vertical Angles

In the figure below, find the measure of each unlabeled angle.

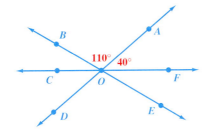

(a) ∠COD
∠COD and ∠AOF are vertical angles so they are congruent. This means they measure the same number of degrees.

The measure of ∠AOF is 40° so the measure of ∠COD is 40° also.

(b) ∠DOE
∠DOE and ∠BOA are vertical angles, so they are congruent.

The measure of ∠BOA is 110° so the measure of ∠DOE is 110° also.

(c) ∠COB
Look at ∠COB, ∠BOA, and ∠AOF. Notice that $\vec{OC}$ and $\vec{OF}$ go in opposite directions. Therefore, ∠COF is a straight angle and measures 180°. To find the measure of ∠COB, subtract the sum of the other two angles from 180°.

$$180° - (110° + 40°) = 180° - (150°) = 30°$$

The measure of ∠COB is 30°.

(d) ∠EOF
∠EOF and ∠COB are vertical angles, so they are congruent. We know from part (c) above that the measure of ∠COB is 30° so the measure of ∠EOF is 30° also.

═══ Work Problem ❼ at the Side.

❼ In the figure below, find the number of degrees in each unlabeled angle.

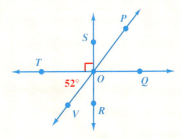

(a) ∠QOR

(b) ∠POQ

(c) ∠VOR

(d) ∠POS

ANSWERS
7. (a) 90° (b) 52° (c) 38° (d) 38°

Focus on Real-Data Applications

The London Eye

The British Airways London Eye was built as a symbol of the turn of the century and represents the cycle of life. It was designed by architects David Marks and Julia Barfield to suspend over the River Thames in Jubilee Gardens, on the bank opposite the Houses of Parliament and Big Ben. It first opened on December 31, 1999.

The London Eye is the largest observation wheel ever designed. It is a unique form of a Ferris wheel that features 32 oval-shaped capsules, each designed to hold 25 people. The wheel structure was assembled on platforms erected in the river bed, and once assembled was lifted into place by a floating crane. The London Eye is the fourth tallest structure in London (and the only one of the four that is open to the public), and the capsules were designed so that passengers have an unobstructed view throughout the ride. On a clear day, passengers can see 25 miles to the 900-year-old Windsor Castle, the world's largest royal residence, used by Queen Elizabeth II.

Another unique feature of the London Eye is that it is designed to rotate continuously at walking speed. At one end of the loading platform, passengers walk out of the capsules, and at the other end of the platform, passengers walk into the capsules. Because the wheel rotates so slowly, the passengers can stand and walk around the capsules during their ride, and enjoy a 360° view of the London skyline. The wheel does stop to allow people in wheelchairs to board. (*Source: The Bankside Press*, 2001.)

Specifications for the London Eye	
Diameter	135 meters
Speed	0.26 meters per second
Time to revolve	30 minutes

1. (a) What is the diameter of the wheel in feet? (*Hint:* Use 1 m ≈ 3.28 ft.)
   (b) The circumference of the wheel is the distance around the outside rim. Find the circumference by multiplying the diameter times 3.1416 and rounding the answer to the nearest foot. (See Section 8.6 for more information about circumference.)

2. Find the distance, to the nearest tenth of a foot, along the rim arc between two adjacent capsules. Recall that there are 32 capsules.

3. If the London Eye is operating at full capacity, how many people will be riding in a 90° section of the wheel? in a 180° section? in a 360° section?

4. If the London Eye operates from 8:00 A.M. until 6:00 P.M. at full capacity, how many people per day can ride?

5. The speed is given as 0.26 meters per second. Find the speed of the London Eye in feet per second. Round the answer to the nearest hundredth. Do you believe that the wheel revolves at "walking speed"? (Try it and see!)

8.2 Exercises

Identify each pair of complementary angles. See Example 1.

1.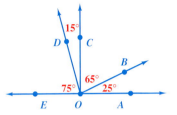

2.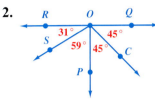

Identify each pair of supplementary angles. See Example 3.

3.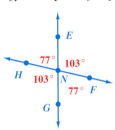

4.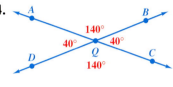

Find the complement of each angle. See Example 2.

5. 40°   6. 35°   7. 86°   8. 59°

Find the supplement of each angle. See Example 4.

9. 130°   10. 75°   11. 90°   12. 5°

In each figure, identify the angles that are congruent. See Example 5.

13.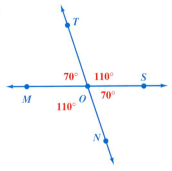

14.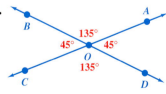

Use your knowledge of vertical angles to answer Exercises 15 and 16. See Examples 6 and 7.

15. In the figure below, ∠AOH measures 37° and ∠COE measures 63°. Find the measure of each of the other angles.

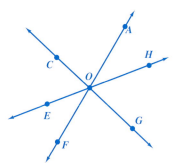

16. In the figure below, ∠POU measures 105° and ∠UOT measures 40°. Find the measure of each of the other angles.

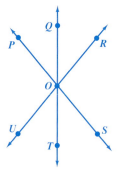

17. In your own words, write a definition of complementary angles and a definition of supplementary angles. Draw a picture to illustrate each definition.

18. Make up a test problem in which a student has to use knowledge of vertical angles. Include a drawing with some angles labeled and ask the student to find the size of the remaining angles. Give the correct answer for your problem.

In each figure, ray AB is parallel to ray CD. Identify two pairs of congruent angles and the number of degrees in each congruent angle.

19.

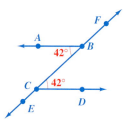

20.

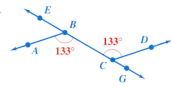

21. Can two obtuse angles be supplementary? Explain why or why not.

22. Can two acute angles be complementary? Explain why or why not.

8.3 RECTANGLES AND SQUARES

A **rectangle** is a figure with four sides that meet to form 90° angles. Each set of opposite sides is parallel and congruent (has the same length).

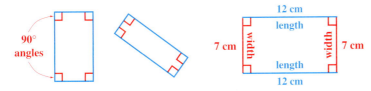

OBJECTIVES

1. Find the perimeter and area of a rectangle.
2. Find the perimeter and area of a square.
3. Find the perimeter and area of a composite shape.

Each longer side of a rectangle is called the length (l) and each shorter side is called the width (w).

Work Problem ❶ **at the Side.**

1 **Find the perimeter and area of a rectangle.** The distance around the outside edges of a figure is the **perimeter** of the figure. Think of how much fence you would need to put around the sides of a garden plot, or how far you would walk if you go around the outside edges of your backyard. In either case you would add up the lengths of the sides. Look at the rectangle above that has the lengths of the sides labeled. To find its perimeter, you add the lengths of the sides.

$$\text{Perimeter} = 12 \text{ cm} + 7 \text{ cm} + 12 \text{ cm} + 7 \text{ cm} = 38 \text{ cm}$$

Because the two long sides are both 12 cm, and the two short sides are both 7 cm, you can also use this formula.

Finding the Perimeter of a Rectangle

Perimeter of a rectangle = (2 • length) + (2 • width)

$$P = 2 \cdot l + 2 \cdot w$$

Example 1 Finding the Perimeter of Rectangles

Find the perimeter of each rectangle.

(a)

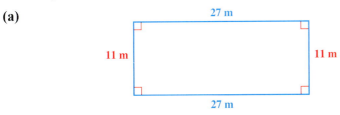

The length is **27 m** and the width is **11 m**.
Use the formula $P = 2 \cdot l + 2 \cdot w$.

$$P = 2 \cdot l + 2 \cdot w$$
$$P = 2 \cdot 27 \text{ m} + 2 \cdot 11 \text{ m}$$
$$P = 54 \text{ m} + 22 \text{ m}$$
$$P = 76 \text{ m}$$

The perimeter of the rectangle (the distance you would walk around the outside edges of the rectangle) is 76 m.

Continued on Next Page

❶ Identify all the rectangles.

(a)

(b)

(c)

(d)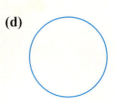

(e)

(f)

(g)

ANSWERS
1. **(a)**, **(b)**, and **(e)** are rectangles; **(c)**, **(d)**, **(f)**, and **(g)** are not.

2 Find the perimeter of each rectangle.

(a)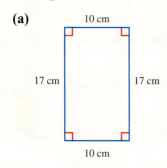

(b)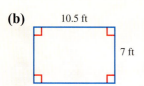

(c) 6 m wide and 11 m long

(d) 0.9 km by 2.8 km

ANSWERS
2. (a) 54 cm (b) 35 ft (c) 34 m
(d) 7.4 km

As a check, you can add up the lengths of the four sides.

$P = 27 \text{ m} + 11 \text{ m} + 27 \text{ m} + 11 \text{ m}$
$P = 76 \text{ m}$ ← Same result as using formula

(b) A rectangle 8.9 m by 12.3 m
You can use the formula, as shown below.

$P = 2 \cdot l + 2 \cdot w$
$P = 2 \cdot 12.3 \text{ m} + 2 \cdot 8.9 \text{ m}$
$P = 24.6 \text{ m} + 17.8 \text{ m}$
$P = 42.4 \text{ m}$

Or, you can add up the lengths of the four sides.

$P = 8.9 \text{ m} + 12.3 \text{ m} + 8.9 \text{ m} + 12.3 \text{ m}$
$P = 42.4 \text{ m}$

Either method will give you the correct result.

Work Problem 2 at the Side.

The *perimeter* of a rectangle is the distance around the *outside edges*. The **area** of a rectangle is the amount of surface *inside* the rectangle. We measure area by seeing how many squares of a certain size are needed to cover the surface inside the rectangle. Think of covering the floor of a rectangular living room with carpet. Carpet is measured in square yards, that is, square pieces that measure 1 yard along each side. Here is a drawing of a living room floor.

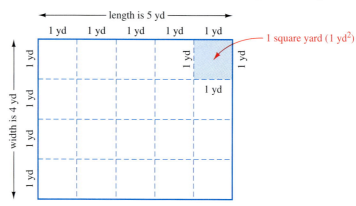

You can see from the drawing that it takes 20 squares to cover the floor. We say that the area of the floor is 20 *square yards*. A shorter way to write square yards is yd^2.

20 **square yards** can be written 20 **yd^2**

To find the number of squares, you can count them, or you can multiply the number of squares in the length (5) times the number of squares in the width (4) to get 20. The formula is given below.

Finding the Area of a Rectangle

Area of a rectangle = length • width
$A = l \cdot w$

Remember to use **square units** when measuring area.

Squares of other sizes can be used to measure area. For smaller areas, you might use these:

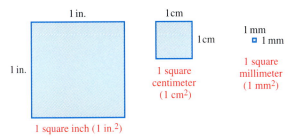

Actual-size drawings

Other sizes of squares that are often used to measure area are listed here, but they are too large to draw on this page.

1 square meter (1 m^2)   1 square foot (1 ft^2)
1 square kilometer (1 km^2)   1 square yard (1 yd^2)
     1 square mile (1 mi^2)

CAUTION

The raised 2 in 4^2 means that you multiply 4 • 4 to get 16. The raised 2 in cm^2 or yd^2 is a short way to write the word *square*. When you see 5 cm^2, say "five square centimeters." Do *not* multiply 5 • 5.

Example 2   Finding the Area of Rectangles

Find the area of each rectangle.

(a)
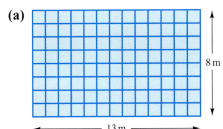

The length of this rectangle is 13 m and the width is 8 m. Use the formula $A = l \cdot w$.

$$A = l \cdot w$$
$$A = 13 \text{ m} \cdot 8 \text{ m}$$
$$A = 104 \text{ square meters}$$

"Square meters" can be written as m^2, so the area is 104 m^2.

(b) A rectangle measuring 7 cm by 21 cm
Use the formula $A = l \cdot w$.

$$A = 21 \text{ cm} \cdot 7 \text{ cm} = 147 \text{ cm}^2$$

The area of the rectangle is 147 cm^2.

3 Find the area of each rectangle.

(a)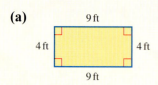

(b) A rectangle that is 6 m long and 0.5 m wide

(c) 8.2 cm by 41.2 cm

CAUTION

The units for *area* will always be *square* units (cm^2, m^2, yd^2, mi^2, and so on). The units for *perimeter* will be cm, m, yd, mi, and so on (not square units).

Work Problem **3** at the Side.

2 **Find the perimeter and area of a square.** A **square** is a rectangle with all sides the same length. Two squares are shown here. Notice the 90° angles.

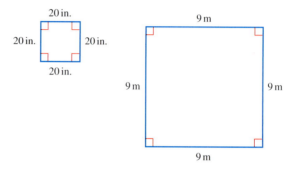

To find the *perimeter* (distance around) of the square on the right, you could add 9 m + 9 m + 9 m + 9 m to get 36 m. A shorter way is to multiply the length of one side times 4, because all four sides are the same length.

Finding the Perimeter of a Square

Perimeter of a square = side + side + side + side

or, $P = 4 \cdot$ side

$P = 4 \cdot s$

As with a rectangle, you can multiply length times width to find the *area* (surface inside) of a square. Because the length and the width are the same in a square, the formula is written as shown below.

Finding the Area of a Square

Area of a square = side • side

$A = s \cdot s$

$A = s^2$

Remember to use *square units* when measuring area.

Example 3 Finding the Perimeter and Area of a Square

(a) Find the perimeter of a square where each side measures 9 m.

Use the formula.

$P = 4 \cdot s$

$P = 4 \cdot 9$ m

$P = 36$ m

Or add up the four sides.

$P = 9$ m $+ 9$ m $+ 9$ m $+ 9$ m

$P = 36$ m

Continued on Next Page

ANSWERS

3. (a) 36 ft^2 (b) 3 m^2 (c) 337.84 cm^2

(b) Find the area of the same square.

$$A = s^2$$
$$A = s \cdot s$$
$$A = 9 \text{ m} \cdot 9 \text{ m}$$
$$A = 81 \text{ m}^2 \qquad \text{Square units for area}$$

CAUTION

Be careful! s^2 means $s \cdot s$. It does **not** mean $2 \cdot s$. In Example 3 above, s is 9 m, so s^2 is 9 m $\cdot$ 9 m = 81 m². It is **not** $2 \cdot 9$ m = 18 m.

Work Problem **4** at the Side.

3 **Find the perimeter and area of a composite shape.** As with any other shape, you can find the perimeter (distance around) an irregular shape by adding up the lengths of the sides. To find the area (surface inside the shape), try to break it up into pieces that are squares or rectangles. Find the area of each piece and then add them together.

Example 4 Finding the Perimeter and Area of a Composite Figure

The floor of a room has the shape shown here.

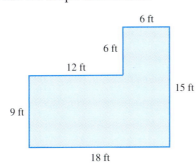

(a) Suppose you want to put a new baseboard (wooden strip) along the base of all the walls. How much material do you need?
Find the perimeter of the room by adding up the lengths of the sides.

$$P = 9 \text{ ft} + 12 \text{ ft} + 6 \text{ ft} + 6 \text{ ft} + 15 \text{ ft} + 18 \text{ ft} = 66 \text{ ft}$$

You need 66 ft of baseboard material.

(b) The carpet you like costs $20.50 per square yard. How much will it cost to carpet the room?
First change the measurements from feet to yards, because the carpet is sold in square yards. There are 3 ft in 1 yd, so multiply by the unit fraction that allows you to divide out feet. For example:

$$\frac{\overset{3}{\cancel{9} \text{ ft}}}{1} \cdot \frac{1 \text{ yd}}{\underset{1}{\cancel{3} \text{ ft}}} = 3 \text{ yd}$$

— Divide out ft.
— Divide out common factors.

Continued on Next Page

4 Find the perimeter and area of each square.

(a)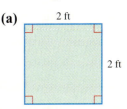
2 ft, 2 ft

(b) 10.5 cm on each side

(c) 2.1 mi on a side

ANSWERS
4. **(a)** $P = 8$ ft; $A = 4$ ft²
(b) $P = 42$ cm; $A = 110.25$ cm²
(c) $P = 8.4$ mi; $A = 4.41$ mi²

5 Carpet costs $19.95 per square yard. Find the cost of carpeting each room. Round your answers to the nearest cent if necessary.

(a)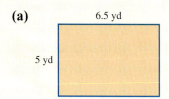

(b)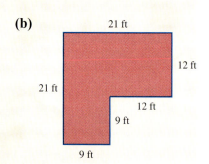

(c) A classroom that is 24 ft long and 18 ft wide

Use the same unit fraction to change the other measurements from feet to yards.

$$\frac{\overset{4}{\cancel{12}\text{ ft}}}{1} \cdot \frac{1\text{ yd}}{\underset{1}{\cancel{3}\text{ ft}}} = 4\text{ yd} \qquad \frac{\overset{2}{\cancel{6}\text{ ft}}}{1} \cdot \frac{1\text{ yd}}{\underset{1}{\cancel{3}\text{ ft}}} = 2\text{ yd}$$

$$\frac{\overset{5}{\cancel{15}\text{ ft}}}{1} \cdot \frac{1\text{ yd}}{\underset{1}{\cancel{3}\text{ ft}}} = 5\text{ yd} \qquad \frac{\overset{6}{\cancel{18}\text{ ft}}}{1} \cdot \frac{1\text{ yd}}{\underset{1}{\cancel{3}\text{ ft}}} = 6\text{ yd}$$

Next, break up the room into two pieces. Use just the measurements for the length and width of each piece.

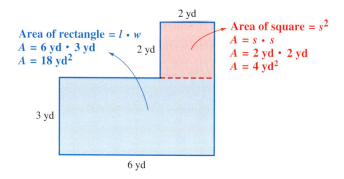

Total area = **18 yd² + 4 yd²** = 22 yd²

Multiply to find the cost of the carpet.

$$\frac{22\text{ yd}^2}{1} \cdot \frac{\$20.50}{1\text{ yd}^2} = \$451.00$$

You could have cut the room into two rectangles. The total area is the same.

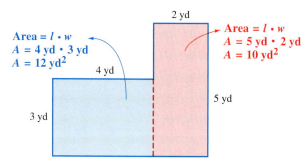

Total area = **12 yd² + 10 yd²** = 22 yd²   Same answer as above

Work Problem 5 at the Side.

ANSWERS

5. (a) 32.5 yd² costs $648.38 (rounded)
(b) 37 yd² costs $738.15
(c) 48 yd² costs $957.60

8.3 Exercises

Find the perimeter and area of each rectangle or square. See Examples 1–3.

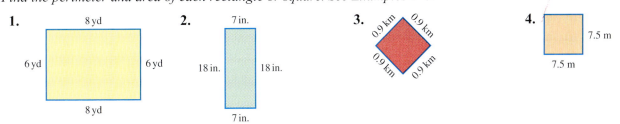

1. 8 yd by 6 yd rectangle
2. 7 in. by 18 in. rectangle
3. square with sides 0.9 km
4. square with sides 7.5 m

Draw a sketch of each square or rectangle and label the lengths of the sides. Then find the perimeter and the area. (Sketches may vary; show your sketches to your instructor.)

5. 10 ft by 10 ft
6. 8 cm by 17 cm
7. 14 m by 0.5 m
8. 2.35 km by 8.4 km

9. A storage building that is 76.1 ft by 22 ft
10. A science lab measuring 12 m by 12 m
11. A square nature preserve 3 mi wide
12. A square of cardboard 20.3 cm on a side

Find the perimeter and area of each figure. See Example 4.

13.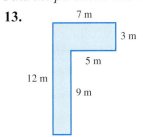

14.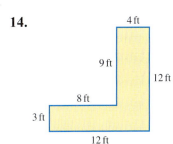

15.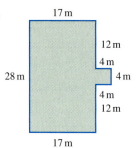

16.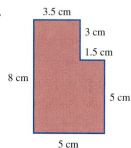

First find the length of the unlabeled side in each figure. Then find the perimeter and area of each figure.

17.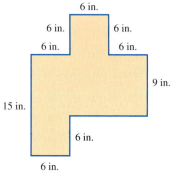

18.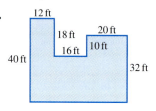

Solve each application problem.

19. The Wang's family room measures 20 ft by 25 ft. They are covering the floor with square tiles that measure 1 ft on a side and cost $0.92 each. How much will they spend on tile?

20. A page in this book measures 27.5 cm from top to bottom and 20.5 cm from side to side. Find the perimeter and the area of the page.

21. Tyra's kitchen is 4.4 m wide and 5.1 m long. She is pasting a decorative border strip that costs $4.99 per meter around the top edge of all the walls. How much will she spend?

22. Mr. and Mrs. Gomez are buying carpet for their square-shaped bedroom that is 5 yd wide. The carpet is $23 per square yard and padding and installation is another $6 per square yard. How much will they spend in all?

23. Advanced Photo System (APS) cameras allow you to choose from three different print sizes each time you snap a photo. The choices are shown below. Find the perimeter and area of each size print. (*Source:* Kodak.)

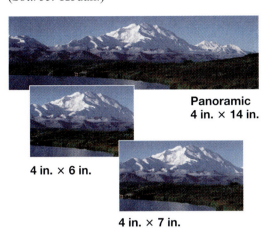

24. The Monterey Bay Aquarium in California lets visitors look into a million-gallon tank through an acrylic panel that is 13 in. thick. The panel is 54 ft long and 15 ft high. What is the perimeter and the area of the panel? (*Source: AAA California Tour Book.*)

25. A regulation football field is 100 yd long (excluding end zones) and has an area of 5300 yd². Find the width of the field. (*Source:* National Football League.)

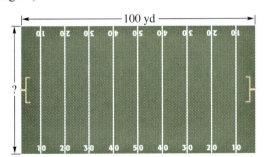

26. There are 14,790 ft² of ice in the rectangular playing area for a major league hockey game (excluding the area behind the goal lines). If the playing area is 85 ft wide, how long is it? (*Source:* National Hockey League.)

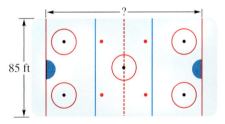

27. A lot is 124 ft by 172 ft. County rules require that nothing be built on land within 12 ft of any edge of the lot. First, add labels to the sketch of the lot, showing the land that cannot be built on. Then find the area of the land that cannot be built on.

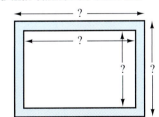

28. Find the cost of fencing needed for this rectangular field. Fencing along the country roads costs $4.25 per foot. Fencing for the other two sides costs $2.75 per foot.

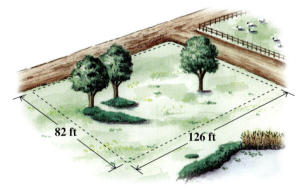

Relating Concepts (Exercises 29–34) For Individual or Group Work

Use your knowledge of perimeter and area to **work Exercises 29–34 in order.**

29. Suppose you have 12 ft of fencing to make a square or rectangular garden plot. Draw sketches of *all* the possible plots that use exactly 12 ft of fencing and label the lengths of the sides. Use only *whole number* lengths. (*Hint:* There are three possibilities.)

30. (a) Find the area of each plot in Exercise 29.

(b) Which plot has the greatest area?

31. Repeat Exercise 29 using 16 ft of fencing. Be sure to draw *all* possible plots that have whole number lengths for the sides.

32. (a) Find the area of each plot in Exercise 31.

(b) Compare your results to those from Exercise 30. What do you notice about the plots with the greatest area?

33. (a) Draw a sketch of a rectangular plot 3 ft by 2 ft. Find the perimeter and area.

(b) Suppose you *double* the length of the plot and *double* the width. Draw a sketch of the enlarged plot and find the perimeter and area.

(c) The *perimeter* of the enlarged plot is how many times greater than the perimeter of the original plot? The *area* of the enlarged plot is how many times greater than the original area?

34. (a) Refer to part (a) of Exercise 33. Suppose you *triple* the length and width of the original plot. Draw a sketch of the enlarged plot and find the perimeter and area.

(b) How many times greater is the *perimeter* of the enlarged plot? How many times greater is the *area* of the enlarged plot?

(c) Suppose you make the length and width *four times greater* in the enlarged plot. What would you predict will happen to the perimeter and area, compared to the original plot?

8.4 PARALLELOGRAMS AND TRAPEZOIDS

A **parallelogram** is a four-sided figure with opposite sides parallel, such as the ones shown below. Notice that opposite sides have the same length.

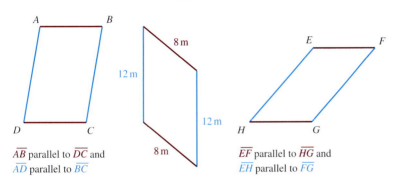

$\overline{AB}$ parallel to $\overline{DC}$ and $\overline{AD}$ parallel to $\overline{BC}$

$\overline{EF}$ parallel to $\overline{HG}$ and $\overline{EH}$ parallel to $\overline{FG}$

OBJECTIVES

1. Find the perimeter and area of a parallelogram.
2. Find the perimeter and area of a trapezoid.

1 Find the perimeter of each parallelogram.

(a)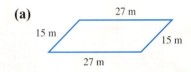

1 **Find the perimeter and area of a parallelogram.** Perimeter is the distance around a figure, so the easiest way to find the perimeter of a parallelogram is to add the lengths of the four sides.

Example 1 Finding the Perimeter of a Parallelogram

Find the perimeter of the middle parallelogram above.

$$P = 12 \text{ m} + 8 \text{ m} + 12 \text{ m} + 8 \text{ m} = 40 \text{ m}$$

Work Problem **1** at the Side.

To find the area of a parallelogram, first draw a dashed line inside the figure as shown here.

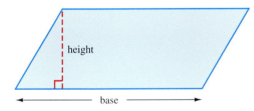

Try this yourself by tracing this parallelogram onto a piece of paper.

The length of the dashed line is the *height* of the parallelogram. It forms a *right angle* with the base. The height is the shortest distance between the base and the opposite side.

Now cut off the triangle created on the left side of the parallelogram and move it to the right side, as shown below.

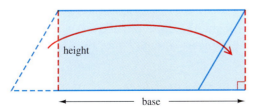

The parallelogram has been made into a rectangle. You can see that the area of the parallelogram and the rectangle are the same. The area of the rectangle is *length* times *width*. In the parallelogram, this translates into *base* times *height*.

(b)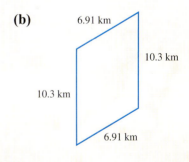

ANSWERS

1. (a) 84 m (b) 34.42 km

546 Chapter 8 Geometry

2 Find the area of each parallelogram.

(a)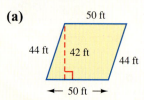

(b)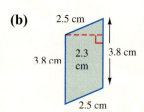

(c) A parallelogram with base $12\frac{1}{2}$ m and height $4\frac{3}{4}$ m (*Hint:* Write $12\frac{1}{2}$ as 12.5 and $4\frac{3}{4}$ as 4.75.)

3 Find the perimeter of each trapezoid.

(a)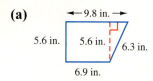

(b)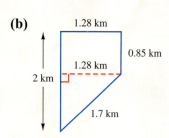

(c) A trapezoid with sides 39.7 cm, 29.2 cm, 74.9 cm, and 16.4 cm

Finding the Area of a Parallelogram

Area of parallelogram = base • height

$$A = b \cdot h$$

Remember to use *square units* when measuring area.

Example 2 Finding the Area of Parallelograms

Find the area of each parallelogram.

(a)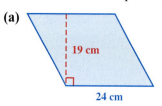

The base is 24 cm and the height is 19 cm. Use the formula $A = b \cdot h$.

$A = b \cdot h$
$A = \text{24 cm} \cdot \text{19 cm}$
$A = 456 \text{ cm}^2$ Square units for area

(b)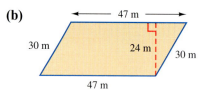

$A = 47 \text{ m} \cdot 24 \text{ m}$
$A = 1128 \text{ m}^2$ Square units for area

Notice that you do *not* use the 30 m sides when finding the area. But you would use them when finding the *perimeter* of the parallelogram.

Work Problem 2 at the Side.

2 Find the perimeter and area of a trapezoid. A **trapezoid** is a four-sided figure with one pair of parallel sides, such as the ones shown below. Opposite sides may *not* have the same length, as in parallelograms.

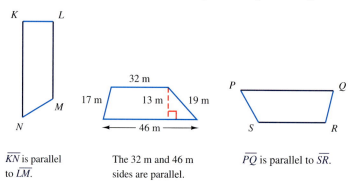

$\overline{KN}$ is parallel to $\overline{LM}$.

The 32 m and 46 m sides are parallel.

$\overline{PQ}$ is parallel to $\overline{SR}$.

Example 3 Finding the Perimeter of a Trapezoid

Find the perimeter of the middle trapezoid above.
You can find the perimeter of any figure by adding the lengths of the sides.

$$P = 17 \text{ m} + 32 \text{ m} + 19 \text{ m} + 46 \text{ m} = 114 \text{ m}$$

Notice that the height (13 m) is *not* part of the perimeter, because the height is *not* one of the *outside edges* of the shape.

Work Problem 3 at the Side.

ANSWERS
2. (a) 2100 ft² (b) 8.74 cm²
   (c) $59\frac{3}{8}$ m² or 59.375 m²
3. (a) 28.6 in. (b) 5.83 km (c) 160.2 cm

Use this formula to find the *area* of a trapezoid.

Finding the Area of a Trapezoid

$$\text{Area} = \frac{1}{2} \cdot \text{height} \cdot (\text{short base} + \text{long base})$$

$$A = \frac{1}{2} \cdot h \cdot (b + B)$$

or $A = 0.5 \cdot h \cdot (b + B)$

Remember to use **square units** when measuring area.

Example 4 Finding the Area of a Trapezoid

Find the area of this trapezoid. The short base and long base are the *parallel* sides.

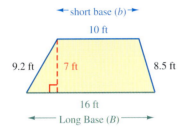

The height (*h*) is **7 ft**, the short base (*b*) is **10 ft**, and the long base (*B*) is **16 ft**. You do *not* need the 9.2 ft or 8.5 ft sides to find the area.

$$A = \frac{1}{2} \cdot h \cdot (b + B)$$

$$A = \frac{1}{2} \cdot 7 \text{ ft} \cdot (10 \text{ ft} + 16 \text{ ft}) \quad \text{Work inside parentheses first.}$$

$$A = \frac{1}{\cancel{2}} \cdot 7 \text{ ft} \cdot (\overset{13}{\cancel{26}} \text{ ft})$$

$A = 91 \text{ ft}^2$ Square units for area

You can also solve the problem by using 0.5, the decimal equivalent for $\frac{1}{2}$, in the formula.

$$A = 0.5 \cdot h \cdot (b + B)$$
$$A = 0.5 \cdot 7 \cdot (10 + 16)$$
$$A = 0.5 \cdot 7 \cdot 26$$
$$A = 91 \text{ ft}^2$$

Calculator Tip Use the parentheses keys on your scientific calculator to work Example 4:

$$0.5 \;\times\; 7 \;\times\; (\; 10 \;+\; 16 \;)\; = \; 91$$

What happens if you do *not* use the parentheses keys? What order of operations will the calculator follow then? (Answer: The calculator will multiply 0.5 times 7 times 10, and then add 16, giving an *incorrect* answer of 51.)

Work Problem ❹ at the Side.

❹ Find the area of each trapezoid.

(a)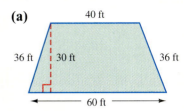

(b)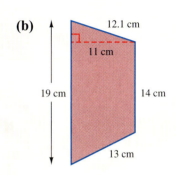

(c) A trapezoid with height 4.7 m, short base 9 m, and long base 10.5 m

ANSWERS
4. (a) 1500 ft² (b) 181.5 cm²
   (c) 45.825 m²

5 Find the area of each floor.

(a)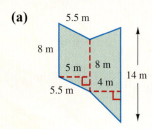

(b)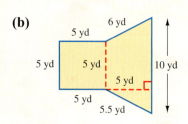

6 Find the cost of carpeting the floors in Problem 5. The cost of carpet is as follows:

(a) Floor (a), $18.50 per square meter.

(b) Floor (b), $28 per square yard.

Example 5   Finding the Area of a Composite Figure

Find the area of this figure.

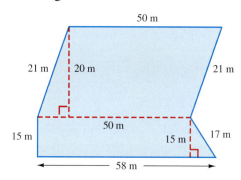

Break the figure into two pieces, a parallelogram and a trapezoid. Find the area of each piece, and then add the areas.

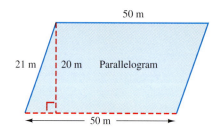

Area of parallelogram
$A = b \cdot h$
$A = 50 \text{ m} \cdot 20 \text{ m}$
$A = 1000 \text{ m}^2$

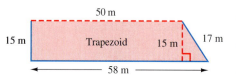

Area of trapezoid
$A = \dfrac{1}{2} \cdot h \cdot (b + B)$
$A = 0.5 \cdot 15 \text{ m} \cdot (50 \text{ m} + 58 \text{ m})$
$A = 810 \text{ m}^2$

Total area = $1000 \text{ m}^2 + 810 \text{ m}^2 = 1810 \text{ m}^2$

Work Problem 5 at the Side.

Example 6   Applying Knowledge of Area

Suppose the figure in Example 5 represents the floor plan of a hotel lobby. What is the cost of labor to install tile on the floor if the labor charge is $35.11 per square meter?

From Example 5, the floor area is 1810 m². To find the labor cost, multiply the number of square meters times the cost of labor per square meter.

$$\text{cost} = \frac{1810 \text{ m}^2}{1} \cdot \frac{\$35.11}{1 \text{ m}^2}$$

$$\text{cost} = \$63{,}549.10$$

The cost of the labor is $63,549.10.

Work Problem 6 at the Side.

ANSWERS
5. (a) 84 m²   (b) 62.5 yd²
6. (a) $1554   (b) $1750

Section 8.4   549

8.4 EXERCISES

Find the perimeter of each figure. See Examples 1 and 3.

1.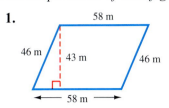

2.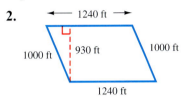

3.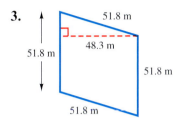

4.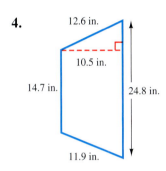

5.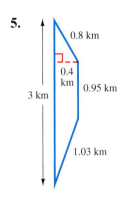

6.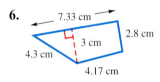

Find the area of each figure. See Examples 2 and 4.

7.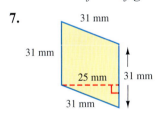

8.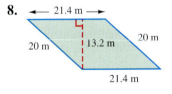

9.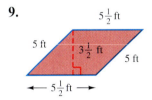

10.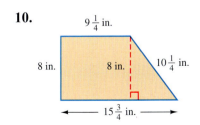

11.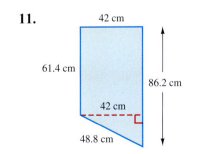

12.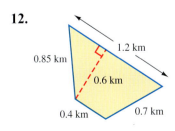

Solve each application problem. First draw a sketch and label the sides and height. (Sketches may vary; show your sketches to your instructor.) See Example 6.

13. The backyard of a new home is shaped like a trapezoid with a height of 45 ft and bases of 80 ft and 110 ft. What is the cost of putting sod on the yard if the landscaper charges $0.33 per square foot for sod?

14. A swimming pool is in the shape of a parallelogram with a height of 9.6 m and base of 12.4 m. Find the labor cost to make a custom solar cover for the pool at a cost of $4.92 per square meter.

15. A piece of fabric for a quilt design is in the shape of a parallelogram. The base is 5 in. and the height is 3.5 in. What is the total area of the 25 parallelogram pieces needed for the quilt?

16. An accountant is paying $832 per month to rent an office in an old building. Her office is shaped like a trapezoid, with bases of 32 ft and 20 ft and a height of 20 ft. How much rent is she paying per square foot?

*Find **two** errors in each student's solution below. Write a sentence explaining each error. Then show how to work the problem correctly.*

17.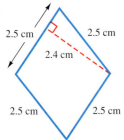

$P = 2.5 \text{ cm} + 2.4 \text{ cm} + 2.5 \text{ cm} + 2.5 \text{ cm} + 2.5 \text{ cm}$

$P = 12.4 \text{ cm}^2$

18.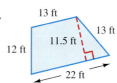

$A = (0.5)(11.5 \text{ ft}) \cdot (12 \text{ ft} + 13 \text{ ft})$

$A = 143.75 \text{ ft}$

 Find the area of each figure. See Example 5.

19.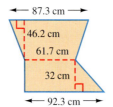

20.

21.

8.5 TRIANGLES

A **triangle** is a figure with exactly three sides. Some examples are shown below.

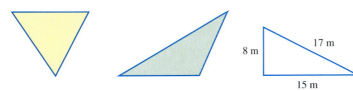

OBJECTIVES

1. Find the perimeter of a triangle.
2. Find the area of a triangle.
3. Given the measures of two angles in a triangle, find the measure of the third angle.

1 Find the perimeter of a triangle. To find the perimeter of a triangle (the distance around the edges), add the lengths of the three sides.

Example 1 — Finding the Perimeter of a Triangle

Find the perimeter of the triangle above on the right.

$$P = 8\text{ m} + 15\text{ m} + 17\text{ m}$$
$$P = 40\text{ m}$$

Work Problem 1 at the Side.

As with parallelograms, you can find the *height* of a triangle by measuring the distance from one corner of the triangle to the opposite side (the base). The height line must be *perpendicular* to the base; that is, it must form a right angle with the base. Sometimes you have to extend the base line in order to draw the height perpendicular to it, as shown on the right in the figures below.

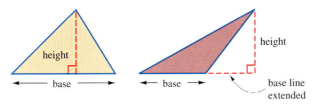

If you cut out two identical triangles and turn one upside down, you can fit them together to form a parallelogram.

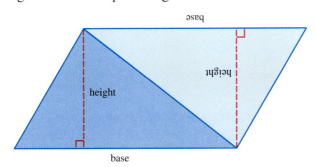

The area of the parallelogram is base times height. Because each triangle is *half* of the parallelogram, the area of one triangle is

$$\frac{1}{2}\text{ of base times height.}$$

1 Find the perimeter of each triangle.

(a)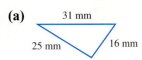

(b)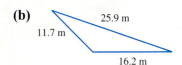

(c) A triangle with sides $6\frac{1}{2}$ yd, $9\frac{3}{4}$ yd, and $11\frac{1}{4}$ yd

ANSWERS
1. (a) 72 mm (b) 53.8 m
   (c) $27\frac{1}{2}$ yd or 27.5 yd

552 Chapter 8 Geometry

➋ Find the area of each triangle.

(a)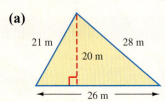

(b)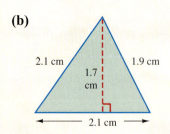

(c)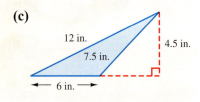

(d)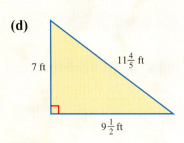

ANSWERS

2. (a) 260 m²   (b) 1.785 cm²
    (c) 13.5 in.²   (d) 33.25 ft² or $33\frac{1}{4}$ ft²

2   **Find the area of a triangle.** Use the following formula to find the *area* of a triangle.

> **Finding the Area of a Triangle**
>
> Area of a triangle = $\frac{1}{2}$ • base • height
>
> $$A = \frac{1}{2} \cdot b \cdot h$$
>
> or $A = 0.5 \cdot b \cdot h$
>
> Remember to use **square units** when measuring area.

Example 2   Finding the Area of Triangles

Find the area of each triangle.

(a)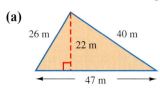

The base is 47 m and the height is 22 m. You do *not* need the 26 m or 40 m sides to find the area.

$$A = \frac{1}{2} \cdot b \cdot h$$

$$A = \frac{1}{\cancel{2}} \cdot 47 \text{ m} \cdot \cancel{22}^{11} \text{ m} \qquad \text{Divide out common factor of 2.}$$

$$A = 517 \text{ m}^2 \qquad \text{Square units for area}$$

(b)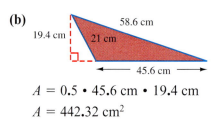

$A = 0.5 \cdot 45.6 \text{ cm} \cdot 19.4 \text{ cm}$

$A = 442.32 \text{ cm}^2$

The base line must be extended to draw the height. However, still use 45.6 cm for *b* in the formula. Because the measurements are decimal numbers, it is easier to use 0.5 (the decimal equivalent of $\frac{1}{2}$) in the formula.

(c)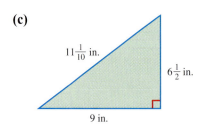

Because two sides of the triangle are perpendicular to each other, use those sides as the base and the height. (Recall that the height line must be perpendicular to the base.)

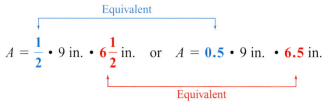

$$A = \frac{1}{2} \cdot 9 \text{ in.} \cdot 6\frac{1}{2} \text{ in.} \quad \text{or} \quad A = 0.5 \cdot 9 \text{ in.} \cdot 6.5 \text{ in.}$$

$$A = 29\frac{1}{4} \text{ in.}^2 \qquad \text{or} \qquad A = 29.25 \text{ in.}^2$$

Work Problem ➋ at the Side.

Example 3 — Using the Concept of Area

Find the area of the shaded part in this figure.

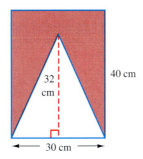

The *entire* figure is a rectangle. Find the area of the rectangle.

$$A = l \cdot w$$
$$A = 30 \text{ cm} \cdot 40 \text{ cm} = 1200 \text{ cm}^2$$

The *un*shaded part is a triangle. Find the area of the triangle.

$$A = \frac{1}{\cancel{2}} \cdot \cancel{30}^{\,15} \text{ cm} \cdot 32 \text{ cm}$$

$$A = 480 \text{ cm}^2$$

Subtract to find the area of the shaded part.

$$\underbrace{A = 1200 \text{ cm}^2}_{\text{Entire area}} - \underbrace{480 \text{ cm}^2}_{\text{Unshaded part}} = \underbrace{720 \text{ cm}^2}_{\text{Shaded part}}$$

Work Problem ③ at the Side.

Example 4 — Applying the Concept of Area

The Department of Transportation cuts triangular signs out of rectangular pieces of metal using the measurements shown above in Example 3. If the metal costs $0.02 per square centimeter, how much does the metal cost for the sign? What is the cost of the metal that is *not* used?

From Example 3, the area of the triangle (the sign) is 480 cm². Multiply that times the cost per square centimeter.

$$\text{cost of sign} = \frac{480 \text{ cm}^2}{1} \cdot \frac{\$0.02}{1 \text{ cm}^2} = \$9.60$$

The metal that is *not* used is the *shaded* part from Example 3. The area is 720 cm².

$$\text{cost of unused metal} = \frac{720 \text{ cm}^2}{1} \cdot \frac{\$0.02}{1 \text{ cm}^2} = \$14.40$$

Work Problem ④ at the Side.

---

❸ Find the area of the shaded part in this figure.

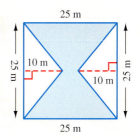

❹ Suppose the figure in Problem 3 above is an auditorium floor plan. The shaded part will be covered with carpet costing $27 per square meter. The rest will be covered with vinyl floor covering costing $18 per square meter. What is the total cost of covering the floor?

Answers
3. 625 m² − 125 m² − 125 m² = 375 m²
4. $10,125 + $2250 + $2250 = $14,625

5 Find the number of degrees in the unlabeled angle.

(a)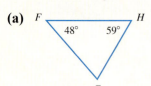

3 **Given the measures of two angles in a triangle, find the measure of the third angle.** The *tri* in *tri*angle means *three*. So the name tells you that a triangle has three angles. The sum of the measures of the three angles in any triangle is *always* 180° (a straight angle). You can see it by drawing a triangle, cutting off the three angles, and rearranging them to make a straight angle.

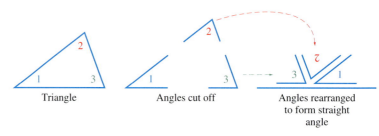

Finding the Unknown Angle Measurement in a Triangle

Step 1 Add the number of degrees in the two angles you are given.

Step 2 Subtract the sum from 180°.

(b)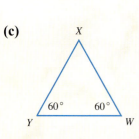

Example 5 Finding an Angle Measurement in Triangles

Find the number of degrees in each indicated angle.

(a) Angle R

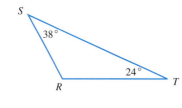

Step 1 Add the two angle measurements you are given.

$$38° + 24° = 62°$$

Step 2 Subtract the sum from 180°.

$$180° - 62° = 118°$$

∠R measures 118°.

(b) Angle F

∠E is a right angle, so it measures 90°.

Step 1 90° + 45° = 135°

Step 2 180° − 135° = 45°

∠F measures 45°.

(c) [triangle XYW with 60° at Y and 60° at W]

Work Problem 5 at the Side.

ANSWERS
5. (a) 73° (b) 35° (c) 60°

8.5 Exercises

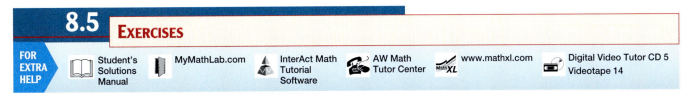

Find the perimeter and area of each triangle. See Examples 1 and 2.

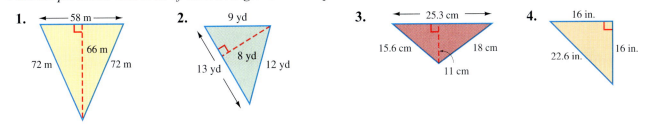

1. 58 m, 66 m, 72 m, 72 m
2. 9 yd, 8 yd, 13 yd, 12 yd
3. 25.3 cm, 15.6 cm, 18 cm, 11 cm
4. 16 in., 22.6 in., 16 in.

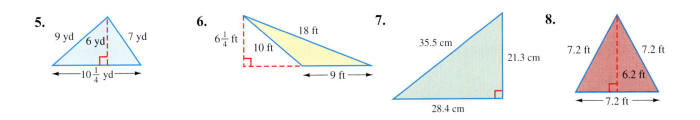

5. 9 yd, 6 yd, 7 yd, $10\frac{1}{4}$ yd
6. $6\frac{1}{4}$ ft, 10 ft, 18 ft, 9 ft
7. 35.5 cm, 21.3 cm, 28.4 cm
8. 7.2 ft, 7.2 ft, 6.2 ft, 7.2 ft

Find the shaded area in each figure. See Example 3.

9.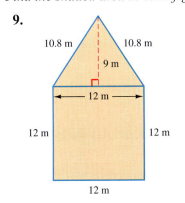

10.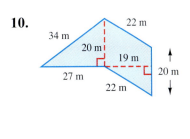

11.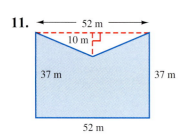

12.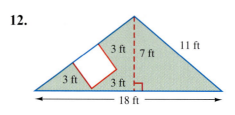

Find the number of degrees in the unlabeled angle. See Example 5.

13.

14.

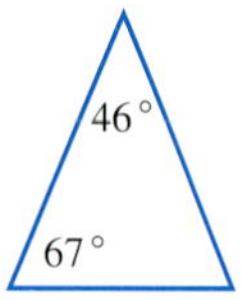

15.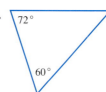

16.

17. Can a triangle have two right angles? Explain your answer.

18. In your own words, explain where the $\frac{1}{2}$ comes from in the formula for area of a triangle. Draw a sketch to illustrate your explanation.

Solve each application problem. See Example 4.

19. A triangular tent flap measures $3\frac{1}{2}$ ft along the base and has a height of $4\frac{1}{2}$ ft. How much canvas is needed to make the flap?

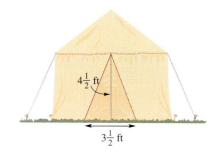

20. A wooden sign in the shape of a right triangle has perpendicular sides measuring 1.5 m and 1.2 m. How much surface area does the sign have?

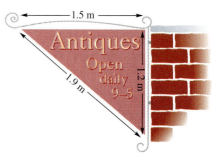

21. A triangular space between three streets has the measurements shown. How much new curbing will be needed to go around the space? How much sod will be needed to cover the space?

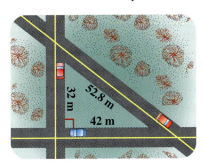

22. Each gable end of a new house has a span of 36 ft and a rise of 9.5 ft. What is the total area of both gable ends of the house?

23. (a) Find the area of one side of the house.
(b) Find the area of one roof section.

All sides of the house are congruent and all roof sections are congruent.

24. The sketch shows the plan for an office building. The shaded area will be a parking lot. What is the cost of building the parking lot if the contractor charges $28.00 per square yard for materials and labor?

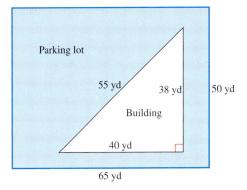

Focus on Real-Data Applications

Interior Design

Suppose that you have just bought a small waterfront house and you plan to remodel the interior by installing tile and carpet on the floors. A sketch of the floor plan is shown below. All measurements are in feet and represent the measurements rounded up to the nearest foot.

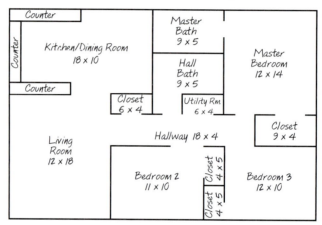

1. How many square feet of floor space are in each of the following rooms?

   (a) Kitchen/dining room

   (b) Living room

   (c) Master bath, and hall bath (including entry) combined

   (d) Hallway, hall closet, and utility room combined

2. Suppose you plan to install floor tiles in the kitchen/dining room, living room, both baths, hall bath entry, hallway, hall closet, and utility room. The flooring company salesperson recommends that you increase your square footage requirement by 5% to compensate for waste. How many square feet of tile are needed?

3. Tiles are sold in boxes of twelve 1-foot square tiles, and partial boxes are not sold. The tile that you selected costs $5.75 per square foot, installed, based on the total number of tiles purchased. The sales tax rate is 8.25%.

   (a) How many boxes of tile must you purchase?

   (b) How much will it cost to install the tile, including sales tax?

4. How many square feet of floor is in all the bedrooms and bedroom closets, combined?

5. Suppose you plan to carpet the bedrooms and closets. The carpet costs $24.95 per square *yard*, installed, and the sales tax rate is 8.25%. The carpet is sold in rolls that are 12 ft wide, so a 3 ft length equals 4 sq yd. (Why?) The salesperson explains that the carpet nap must run in the same direction across carpet seams and recommends that you purchase a 45 ft length of 12-ft-wide carpet.

   (a) Write how the salesperson may have computed the 45 ft. length. The 45 ft length is how many square *yards* of carpet?

   (b) A common industry formula is to compute $(L \times W)/8$, rounded up to the next square yard, to estimate the number of square yards. How accurate is that formula for this job?

   (c) How much will it cost to install the carpet, including sales tax?

8.6 CIRCLES

1 **Find the radius and diameter of a circle.** Suppose you start with one dot on a piece of paper. Then you draw a bunch of dots that are each 2 cm away from the first dot. If you draw enough dots (points) you'll end up with a *circle*. Each point on the circle is exactly 2 cm away from the *center* of the circle. The 2 cm distance is called the *radius*, *r*, of the circle. The distance across the circle (passing through the center) is called the *diameter*, *d*, of the circle.

OBJECTIVES

1 Find the radius and diameter of a circle.

2 Find the circumference of a circle.

3 Find the area of a circle.

4 Become familiar with Latin and Greek prefixes used in math terminology.

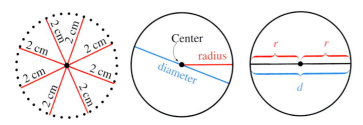

Circle, Radius, and Diameter

A **circle** is a two-dimensional (flat) figure with all points the same distance from a fixed center point.

The **radius** (*r*) is the distance from the center of the circle to any point on the circle.

The **diameter** (*d*) is the distance across the circle passing through the center.

The circle on the right above illustrates the relationships between the radius and diameter. Use the formulas below.

Finding the Diameter and Radius of a Circle

$$\text{diameter} = 2 \cdot \text{radius}$$
$$d = 2 \cdot r$$

and $\quad r = \dfrac{d}{2}$

Example 1   Finding the Diameter and Radius of Circles

Find the unknown length of the diameter or radius in each circle.

(a)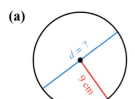

Because the radius is 9 cm, the diameter is twice as long.

$$d = 2 \cdot r$$
$$d = 2 \cdot 9 \text{ cm}$$
$$d = 18 \text{ cm}$$

Continued on Next Page

1 Find the unknown length of the diameter or radius in each circle.

(a)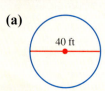

(b)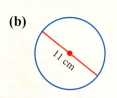

(c)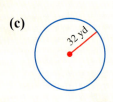

(d)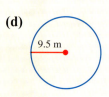

(b)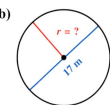

The radius is half the diameter.

$$r = \frac{d}{2}$$

$$r = \frac{17 \text{ m}}{2}$$

$$r = 8.5 \text{ m} \quad \text{or} \quad 8\frac{1}{2} \text{ m}$$

Work Problem 1 at the Side.

2 Find the Circumference of a circle. The perimeter of a circle is called its **circumference**. Circumference is the distance around the edge of a circle.

The diameter of the can in the drawing is about 10.6 cm, and the circumference of the can is about 33.3 cm. Dividing the circumference of the circle by the diameter gives an interesting result.

$$\frac{\text{circumference}}{\text{diameter}} = \frac{33.3}{10.6} \approx 3.14 \quad \text{Rounded to the nearest hundredth}$$

Dividing the circumference of *any* circle by its diameter *always* gives an answer close to 3.14. This means that going around the edge of any circle is a little more than 3 times as far as going straight across the circle.

This ratio of circumference to diameter is called π (the Greek letter **pi**, pronounced PIE). There is no decimal that is exactly equal to π, but here is the *approximate* value.

$$\pi \approx 3.14159265359$$

Rounding the Value of Pi (π)

We usually round π to 3.14. Therefore, calculations involving π will give approximate answers and should be written using the $\approx$ symbol.

Use the following formulas to find the *circumference* of a circle.

Finding the Circumference (Distance around a Circle)

$$\text{circumference} = \pi \cdot \text{diameter}$$
$$C = \pi \cdot d$$

or, because $d = 2 \cdot r$, then $C = \pi \cdot 2 \cdot r$ usually written $C = 2 \cdot \pi \cdot r$

ANSWERS
1. (a) $r = 20$ ft (b) $r = 5.5$ cm
   (c) $d = 64$ yd (d) $d = 19$ m

Section 8.6 Circles   561

Example 2 Finding the Circumference of Circles

Find the circumference of each circle. Use 3.14 as the approximate value for π. Round answers to the nearest tenth.

(a)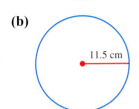

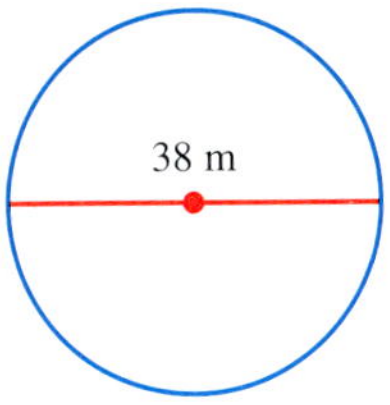

The diameter is 38 m, so use the formula with d in it.

$C = \pi \cdot d$
$C \approx 3.14 \cdot 38$ m
$C \approx 119.3$ m   Rounded

(b)

In this example, the radius is labeled, so it is easier to use the formula with r in it.

$C = 2 \cdot \pi \cdot r$
$C \approx 2 \cdot 3.14 \cdot 11.5$ cm
$C \approx 72.2$ cm   Rounded

Calculator Tip Scientific calculators have a π key. Try pressing it. With a 10-digit display, you'll see the value of π to the nearest billionth.

3.141592654

But this is still an approximate value, although it is more precise than rounding π to 3.14. Try finding the circumference in Example 2(a) above using the π key.

π × 38 = 119.3805208   Rounds to 119.4

When you used 3.14 as the approximate value of π, the result rounded to 119.3, so the answers are slightly different. In this book we will use 3.14 instead of the π key. Our measurements of radius and diameter are given as whole numbers or with tenths, so it is acceptable to round π to hundredths. And, some students may be using standard calculators without a π key, or doing the calculations by hand.

Work Problem ❷ at the Side.

3 Find the area of a circle.   To find the formula for the area of a circle, start by cutting two circles into many pie-shaped pieces.

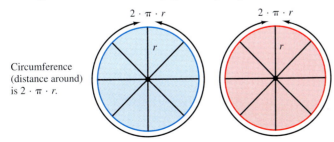

Circumference (distance around) is $2 \cdot \pi \cdot r$.

Unfold the circles, much as you might "unfold" a peeled orange, and put them together as shown here.

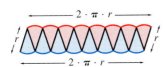

❷ Find the circumference of each circle. Use 3.14 as the approximate value for π. Round answers to the nearest tenth.

(a)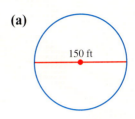

(b)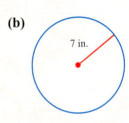

(c) diameter 0.9 km

(d) radius 4.6 m

ANSWERS
2. (a) $C \approx 471$ ft (b) $C \approx 44.0$ in.
   (c) $C \approx 2.8$ km (d) $C \approx 28.9$ m

562 Chapter 8 Geometry

❸ Find the area of each circle. Use 3.14 for π. Round your answers to the nearest tenth.

(a)
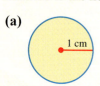

(b)
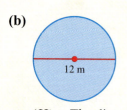

(*Hint:* The diameter is 12 m so $r =$ _____ m)

🖩 (c)
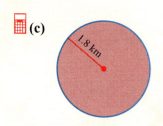

🖩 (d)
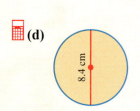

ANSWERS
3. (a) $A \approx 3.1$ cm²   (b) $A \approx 113.0$ m²
    (c) $A \approx 10.2$ km²   (d) $A \approx 55.4$ cm²

The figure is approximately a rectangle with width r (the radius of the original circle) and length $2 \cdot \pi \cdot r$ (the circumference of the original circle). The area of the "rectangle" is length times width.

$$\text{Area} = l \cdot w$$
$$\text{Area} = 2 \cdot \pi \cdot \underbrace{r \cdot r}$$
$$\text{Area} = 2 \cdot \pi \cdot r^2 \quad \leftarrow \text{Recall that } r \cdot r \text{ is } r^2$$

Because the "rectangle" was formed from *two* circles, the area of *one* circle is half as much.

$$\frac{1}{\cancel{2}} \cdot \cancel{2} \cdot \pi \cdot r^2 = 1 \cdot \pi \cdot r^2 \quad \text{or simply} \quad \pi \cdot r^2$$

Finding the Area of a Circle

Area of a circle = π • radius • radius
$$A = \pi \cdot r^2$$

Remember to use *square units* when measuring area.

Example 3   Finding the Area of Circles

Find the area of each circle. Use 3.14 for π. Round your answers to the nearest tenth.

(a) A circle with radius 8.2 cm

Use the formula $A = \pi \cdot r^2$, which means $\pi \cdot r \cdot r$.

$$A = \pi \cdot r \cdot r$$
$$A \approx 3.14 \cdot 8.2 \text{ cm} \cdot 8.2 \text{ cm}$$
$$A \approx 211.1 \text{ cm}^2 \quad \text{Rounded; square units for area}$$

(b)

To use the formula, you need to know the radius (r). In this circle, the diameter is 10 ft. First find the radius.

$$r = \frac{d}{2}$$
$$r = \frac{10 \text{ ft}}{2} = 5 \text{ ft}$$

Now find the area.

$$A \approx 3.14 \cdot 5 \text{ ft} \cdot 5 \text{ ft}$$
$$A \approx 78.5 \text{ ft}^2 \quad \text{Square units for area}$$

CAUTION

When finding *circumference*, you can start with either the radius or the diameter. When finding *area*, you must use the *radius*. If you are given the diameter, divide it by 2 to find the radius. Then find the area.

Work Problem ❸ **at the Side.**

Section 8.6 Circles 563

Calculator Tip You can find the area of the circle in Example 3(a) on your calculator. The first method works on both scientific and standard calculators:

$$3.14 \times 8.2 \times 8.2 = 211.1336$$

You round the answer to 211.1 (nearest tenth).

On a scientific calculator you can also use the x^2 key, which automatically squares the number you enter (that is, multiplies the number times itself):

$$3.14 \times 8.2 \; x^2 \; \underline{67.24} = 211.1336$$

Appears automatically;
8.2 × 8.2 is 67.24

In the next example we find the area of a *semicircle,* which is half the area of a circle.

Example 4 Finding the Area of a Semicircle

Find the area of the semicircle. Use 3.14 for π. Round your answer to the nearest tenth.

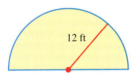

First, find the area of an entire circle with a radius of 12 ft.

$$A = \pi \cdot r \cdot r$$
$$A \approx 3.14 \cdot 12 \text{ ft} \cdot 12 \text{ ft}$$
$$A \approx 452.16 \text{ ft}^2 \quad \text{Do not round yet.}$$

Divide the area of the whole circle by 2 to find the area of the semicircle.

$$\frac{452.16 \text{ ft}^2}{2} = 226.08 \text{ ft}^2$$

The *last* step is rounding 226.08 to the nearest tenth.

$$\text{Area of semicircle} \approx 226.1 \text{ ft}^2 \quad \text{Rounded}$$

Work Problem 4 at the Side.

Example 5 Applying the Concept of Circumference

A circular rug is 8 ft in diameter. The cost of fringe for the edge is $2.25 per foot. What will it cost to add fringe to the rug? Use 3.14 for π.

$$\text{Circumference} = \pi \cdot d$$
$$C \approx 3.14 \cdot 8 \text{ ft}$$
$$C \approx 25.12 \text{ ft}$$

$$\text{cost} = \text{cost per foot} \cdot \text{circumference}$$
$$\text{cost} = \frac{\$2.25}{1 \text{ ft}} \cdot \frac{25.12 \text{ ft}}{1}$$
$$\text{cost} = \$56.52$$

Work Problem 5 at the Side.

4 Find the area of each semicircle. Use 3.14 for π. Round your answers to the nearest tenth.

(a)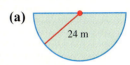

(b)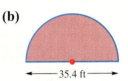

(c)

5 Find the cost of binding around the edge of a circular rug that is 3 m in diameter. The binder charges $4.50 per meter. Use 3.14 for π.

Answers
4. (a) $A \approx 904.3 \text{ m}^2$ (b) $A \approx 491.9 \text{ ft}^2$
  (c) $A \approx 150.8 \text{ m}^2$
5. $42.39

6 Find the cost of covering the underside of the rug in Problem 5 with a nonslip rubber backing. The rubber backing costs $2 per square meter.

7 (a) Here are some more prefixes you have seen in this textbook. List at least one math term and one nonmathematical word that use each prefix.

dia (through):

fract (break):

par (beside):

per (divide):

peri (around):

rad (ray):

rect (right):

sub (below):

(b) How could you use your knowledge of prefixes to remember the difference between perimeter and area?

ANSWERS
6. $14.13
7. (a) Some possibilities are:
   diameter; diagonal
   fraction; fracture
   parallel; paramedic
   percent; per capita
   perimeter; periscope
   radius; radiate
   rectangle; rectify
   subtract; submarine.
   (b) *Peri* in perimeter means "around," so perimeter is the distance *around* the edges of a shape.

Example 6 Applying the Concept of Area

Find the cost of covering the rug in Example 5 with a plastic cover. The material for the cover costs $1.50 per square foot. Use 3.14 for π.

First find the radius.

$$r = \frac{d}{2} = \frac{8 \text{ ft}}{2} = 4 \text{ ft}$$

Then, $A = \pi \cdot r^2$

$A \approx 3.14 \cdot 4 \text{ ft} \cdot 4 \text{ ft}$

$A \approx 50.24 \text{ ft}^2$

$$\text{cost} = \frac{\$1.50}{1 \text{ ft}^2} \cdot \frac{50.24 \text{ ft}^2}{1} = \$75.36$$

Work Problem 6 at the Side.

4 Become familiar with Latin and Greek prefixes used in math terminology. Many English words are built from Latin or Greek root words and prefixes. Knowing the meaning of the more common ones can help you figure out the meaning of terms in many subject areas, including math.

Example 7 Using Prefixes to Understand Math Terms

(a) Listed below are some Latin and Greek root words and prefixes with their meanings in parentheses. You've already seen math terms in this textbook that use these prefixes. List at least one math term and one nonmathematical word that use each prefix or root word.

cent (100): **cent**imeter; **cent**ury

circum (around): **circum**ference; **circum**vent

de (down): **de**nominator; **de**duction

dec (10): **dec**imal; **Dec**ember (originally the 10th month in the old calendar)

There are many answers. These are some of the possibilities.

(b) Suppose you have trouble remembering which part of a fraction is the denominator. How could your knowledge of prefixes help in this situation?

The *de* prefix in *denominator* means "down" so the denominator is the number *down* below the fraction bar.

Work Problem 7 at the Side.

NOTE

Here are some additional prefixes and root words and their meanings that you will see in the rest of Chapter 8, in Chapter 9, and in other math classes. An example of a math term and a nonmathematical word are shown for each one.

equ (equal): **equ**ation; **equ**inox

hemi (half): **hemi**sphere; **hemi**trope

lateral (side): quadri**lateral**; bi**lateral**

re (back or again): **re**ciprocal; **re**duce

Section 8.6   565

8.6 EXERCISES

FOR EXTRA HELP: Student's Solutions Manual, MyMathLab.com, InterAct Math Tutorial Software, AW Math Tutor Center, www.mathxl.com, Digital Video Tutor CD 5 Videotape 14

Find the unknown length in each circle. See Example 1.

1. 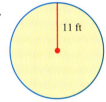 2. 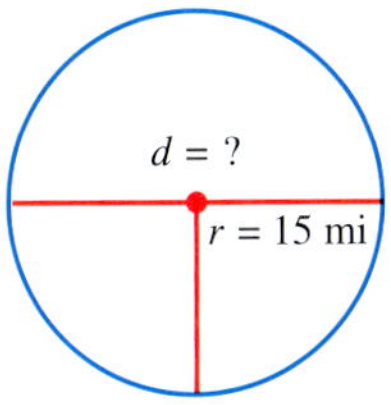 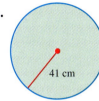 3. 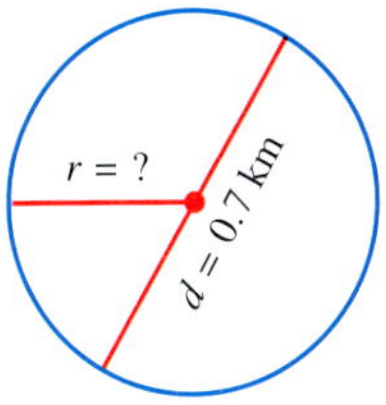 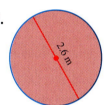 4.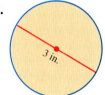

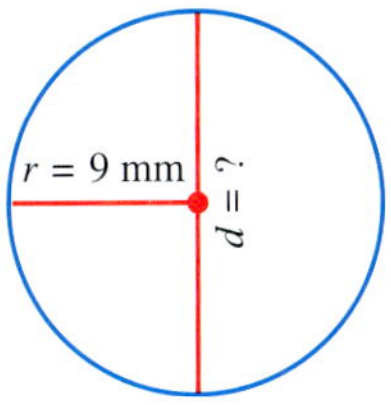

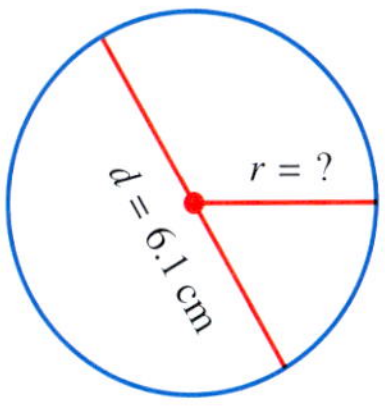

Find the circumference and area of each circle. Use 3.14 as the approximate value for π. Round your answers to the nearest tenth. See Examples 2 and 3.

5. (11 ft)   6. (41 cm)   7. (2.6 m)   8. (3 in.)

Find the circumference and area of circles having the following diameters. Use 3.14 for π. Round your answers to the nearest tenth. See Examples 2 and 3.

9. $d = 15$ cm

10. $d = 39$ ft

11. $d = 7\frac{1}{2}$ ft

12. $d = 4\frac{1}{2}$ yd

13. $d = 8.65$ km

14. $d = 19.5$ mm

Find each shaded area. Note that Exercises 15 and 18 contain semicircles. Use 3.14 as the approximate value of π. Round your answers to the nearest tenth if necessary. See Example 4.

15. 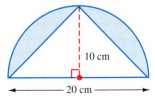   16.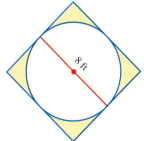

17.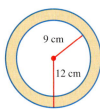

18.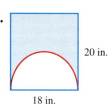

19. How would you explain π to a friend who is not in your math class? Write an explanation. Then make up a test question that requires the use of π, and show how to solve it.

20. Explain how circumference and perimeter are alike. How are they different? Make up two problems, one involving perimeter, the other circumference. Show how to solve your problems.

Solve each application problem. Use 3.14 as the approximate value of π. Round your answers to the nearest tenth. See Examples 5 and 6.

21. How far does a point on the tread of a tire move in one turn if the diameter of the tire is 70 cm?

22. If you swing a ball held at the end of a string 2 m long, how far will the ball travel on each turn?

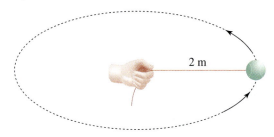

23. A wave energy extraction device is a huge undersea dome used to harness the power of ocean waves. The base of the dome has a radius of 125 ft. Find its circumference.

24. Find the area of the base of the dome in Exercise 23.

For Exercises 25–28, first draw a circle and label the radius or diameter. Then solve the problem. (Sketches may vary; show your sketches to your instructor.)

25. A radio station can be heard 150 miles in all directions during evening hours. How many square miles are in the station's broadcast area?

26. An earthquake was felt by people 900 km away in all directions from the epicenter (the source of the earthquake). How much area was affected by the quake?

27. The diameter of Diana Hestwood's wristwatch is 1 in. and the radius of the clock face on her kitchen wall is 3 in. Find the circumference and the area of each clock face.

28. The diameter of the largest known ball of twine is 12 ft 9 in. (*Source: Guinness Book of Records.*) The sign posted near the ball says it has a circumference of 40 ft. Is the sign correct? (*Hint:* First change 9 in. to feet and add it to 12 ft.)

29. On the first page of this chapter, you read about a forester measuring the circumferences of trees. If the circumference of one tree is 144 cm, what is the diameter of the tree?

30. In Atlanta, Interstate 285 circles the city and is known as the "perimeter." If the circumference of the circle made by the highway is 62.8 miles, find

(a) the diameter of the circle, and

(b) the area inside the circle.

(*Source: Greater Atlanta Newcomer's Guide.*)

31. Find the cost of sod, at $1.76 per square foot, for this playing field that has a semicircle on each end.

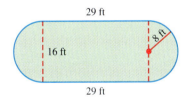

32. Find the area of this skating rink.

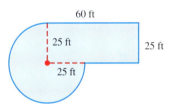

Use the information about prefixes in Example 7 to answer Exercises 33 and 34.

33. Explain how you could use the information about prefixes to remember the difference between radius, diameter, and circumference.

34. Explain how you could use the information about prefixes to avoid confusion between parallel and perpendicular lines.

Relating Concepts (Exercises 35–40) For Individual or Group Work

 Use the table below to **work Exercises 35–40 in order.**

Find the best buy for each type of pizza. The best buy is the lowest cost per square inch of pizza. All the pizzas are circular in shape, and the measurement given on the menu board is the diameter of the pizza in inches. Use 3.14 as the approximate value of π. Round the area to the nearest tenth. Round cost per square inch to the nearest thousandth.

Pizza Menu	Small $7\frac{1}{2}$"	Medium 13"	Large 16"
Cheese only	$2.80	$ 6.50	$ 9.30
"The Works"	$3.70	$ 8.95	$14.30
Deep-dish combo	$4.35	$10.95	$15.65

35. Find the area of a small pizza.

36. Find the area of a medium pizza.

37. Find the area of a large pizza.

38. What is the cost per square inch for each size of cheese pizza? Which size is the best buy?

39. What is the cost per square inch for each size of "The Works" pizza? Which size is the best buy?

40. You have a coupon for 95¢ off any small pizza. What is the cost per square inch for each size of deep dish combo pizza? Which size is the best buy?

Summary Exercises on PERIMETER, CIRCUMFERENCE, AND AREA

1. Draw a sketch of each of these shapes: **(a)** square, **(b)** rectangle, **(c)** parallelogram. On each sketch, indicate 90° angles and show which sides are the same length.

2. **(a)** Draw a sketch of a circle and show the radius.

   (b) Draw another circle and show the diameter.

   (c) Describe the relationship between the radius and diameter of a circle.

3. Describe how you can find the perimeter of any flat shape with straight sides.

4. In your own words, describe the difference between finding the perimeter of a shape and finding the area of the shape.

5. Match each shape to its corresponding area formula.

   Shapes   **Area Formulas**

   parallelogram _____   **(a)** $A = \frac{1}{2} \cdot b \cdot h$

   square _____   **(b)** $A = \pi \cdot r^2$

   trapezoid _____   **(c)** $A = l \cdot w$

   circle _____   **(d)** $A = \frac{1}{2} \cdot h \cdot (b + B)$

   rectangle _____   **(e)** $A = s^2$

   triangle _____   **(f)** $A = b \cdot h$

6. **(a)** If you know the *radius* of a circle, which formula do you use to find its circumference?

   (b) If you know the *diameter* of a circle, which formula do you use to find its circumference?

   (c) If you know the *diameter* of a circle, what must you do *before* using the formula for finding the area?

7. Find the perimeter and area of each triangle, to the nearest tenth.

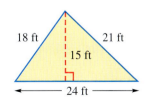
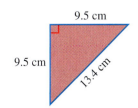

8. Some answers from a student's test paper are listed below. The *number* part of each answer is correct, but the *units* are not. Rewrite each answer with the correct units.

   (a) $A = 12$ cm

   (b) $P = 6\frac{1}{2}$ ft²

   (c) $C \approx 28.5$ m²

   (d) $A = 307$ in.

Find the perimeter and area of each shape. Round answers to the nearest tenth.

9.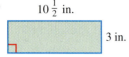

10.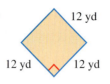

11.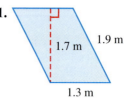

12.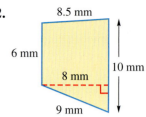

For each circle, find (a) the diameter or radius, (b) the circumference, and (c) the area. Use 3.14 as the approximate value of π. Round answers to the nearest tenth.

13.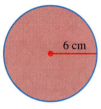

14.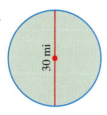

15.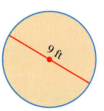

Find the area of each shaded region. Use 3.14 as the approximate value of π. Round answers to the nearest tenth.

16.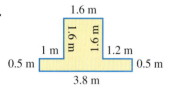

17.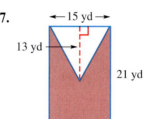

18.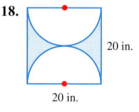

Solve each application problem. Use 3.14 as the approximate value of π. Round answers to the nearest tenth.

19. The Mormons traveled west to Utah by covered wagon in 1847. They tied a rag to a wagon wheel to keep track of the distance they traveled. The radius of the wheel was 2.33 ft. How far did the rag travel each time the wheel made a complete revolution? (*Source:* Trail of Hope.) Bonus question: How many wheel revolutions equaled one mile?

20. The front door for a new log home is 0.9 m wide and 2 m high. How much will it cost for weather strip material to go around all edges of the door, if weather strip costs $0.77 per meter?

8.7 VOLUME

OBJECTIVES

Find the volume of a
1. rectangular solid;
2. sphere;
3. cylinder;
4. cone and pyramid.

1 Find the volume of a rectangular solid. A shoe box and a cereal box are examples of three-dimensional (or solid) figures. The three dimensions are length, width, and height. (A rectangle or square is a two-dimensional figure. The two dimensions are length and width.) If you want to know how much the shoe box will hold, you find its *volume*. We measure volume by seeing how many cubes of a certain size will fill the space inside the box. All the edges of a cube have the same length. Three sizes of *cubic units* are shown below.

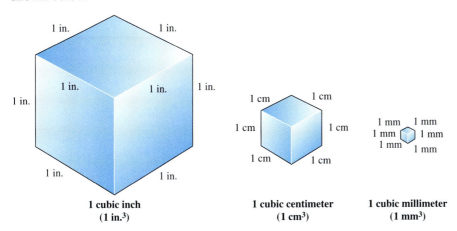

1 cubic inch (1 in.³)   1 cubic centimeter (1 cm³)   1 cubic millimeter (1 mm³)

Some other sizes of cubes that are used to measure volume are 1 cubic foot (1 ft³), 1 cubic yard (1 yd³), and 1 cubic meter (1 m³).

CAUTION

The raised 3 in 4^3 means that you multiply $4 \cdot 4 \cdot 4$ to get 64. The raised 3 in cm³ or ft³ is a short way to write the word *cubic*. When you see 5 cm³, say "five *cubic* centimeters." Do *not* multiply $5 \cdot 5 \cdot 5$.

Volume

Volume is a measure of the space inside a solid shape. The volume of a solid is how many cubic units it takes to fill the solid.

Use the following formula to find the *volume* of rectangular solids (box-like shapes).

Finding the Volume of Rectangular Solids

Volume of rectangular solid = length • width • height
$$V = l \cdot w \cdot h$$

Remember to use ***cubic units*** when measuring volume.

1 Find the volume of each box. Round your answers to the nearest tenth if necessary.

(a)
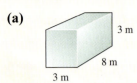

(b)
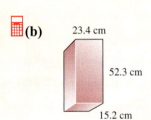

(c) Length $6\frac{1}{4}$ ft, width $3\frac{1}{2}$ ft, height 2 ft

Example 1 Finding the Volume of Rectangular Solids

Find the volume of each box.

(a)
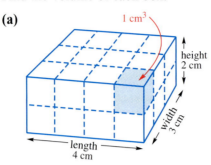

Each cube that fits in the box is 1 cubic centimeter (1 cm³). To find the volume, you can count the number of cubes.

Bottom layer has 12 cubes. } total of 24 cubes (24 cm³)
Top layer has 12 cubes.

Or, you can use the formula for rectangular solids.

$$V = l \cdot w \cdot h$$
$$V = 4 \text{ cm} \cdot 3 \text{ cm} \cdot 2 \text{ cm}$$
$$V = 24 \text{ cm}^3 \quad \text{Cubic units for volume.}$$

(b)
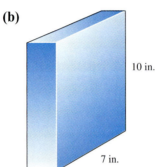

Use the formula.

$$V = 7 \text{ in.} \cdot 2\frac{1}{2} \text{ in.} \cdot 10 \text{ in.}$$

$$V = \frac{7 \text{ in.}}{1} \cdot \frac{\overset{5}{\cancel{5}} \text{ in.}}{\underset{1}{\cancel{2}}} \cdot \frac{\overset{5}{\cancel{10}} \text{ in.}}{1} = 175 \text{ in.}^3$$

If you like, use 2.5 in., the decimal equivalent of $2\frac{1}{2}$ in., for the width.

$$V = 7 \text{ in.} \cdot 2.5 \text{ in.} \cdot 10 \text{ in.} = 175 \text{ in.}^3$$

Work Problem 1 at the Side.

2 **Find the volume of a sphere.** A *sphere* is shown here. Examples of spheres include baseballs, oranges, and the earth. (The last two aren't perfect spheres, but they're close.)

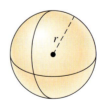

As with circles, the *radius* of a sphere is the distance from the center to the edge of the sphere. Use the following formula to find the *volume* of a sphere.

Finding the Volume of a Sphere

$$\text{Volume of sphere} = \frac{4}{3} \cdot \pi \cdot r \cdot r \cdot r$$

$$V = \frac{4}{3} \cdot \pi \cdot r^3 \quad \text{or} \quad \frac{4 \cdot \pi \cdot r^3}{3}$$

Remember to use *cubic units* when measuring volume.

Answers
1. (a) $V = 72$ m³ (b) $V \approx 18{,}602.1$ cm³
   (c) $V = 43\frac{3}{4}$ ft³ or $V \approx 43.8$ ft³

Section 8.7 Volume **573**

Example 2 Finding the Volume of Spheres

Find the volume of each sphere with the help of a calculator. Use 3.14 as the approximate value of π. Round your answers to the nearest tenth.

 (a)

$$V = \frac{4}{3} \cdot \pi \cdot r^3$$

$$V \approx \frac{4 \cdot 3.14 \cdot 9 \text{ m} \cdot 9 \text{ m} \cdot 9 \text{ m}}{3}$$

$V \approx 3052.08$ *Now round to tenths.*
$V \approx 3052.1 \text{ m}^3$ *Cubic units for volume*

 (b)

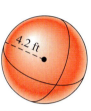

$$V \approx \frac{4 \cdot 3.14 \cdot 4.2 \text{ ft} \cdot 4.2 \text{ ft} \cdot 4.2 \text{ ft}}{3}$$

$V \approx 310.18176$ *Now round to tenths.*
$V \approx 310.2 \text{ ft}^3$ *Cubic units for volume*

Calculator Tip You can find the volume of the sphere in Example 2(b) on your calculator. The first method works on both scientific and standard calculators:

4 ⊗ 3.14 ⊗ 4.2 ⊗ 4.2 ⊗ 4.2 ⊘ 3 ⊜ 310.18176

Round the answer to 310.2 ft³.

On a scientific calculator you can use the y^x key to calculate r^3 (to multiply the radius times itself three times).

4 ⊗ 3.14 ⊗ 4.2 y^x 3 ⊘ 3 ⊜ 310.18176
           $\underbrace{}_{r^3}$

Recall that we are using 3.14 as the approximate value for π instead of using the π key.

You can also use the y^x key with other exponents. For example:

To find 2^5, press 2 y^x 5 ⊜.   Answer is 32.

To find 6^4, press 6 y^x 4 ⊜.   Answer is 1296.

Work Problem ❷ at the Side.

Half a sphere is called a *hemisphere*. The volume of a hemisphere is *half* the volume of a sphere. Use the following formula to find the *volume* of a hemisphere.

Finding the Volume of a Hemisphere

$$V = \frac{2}{3} \cdot \pi \cdot r^3 \quad \text{or} \quad \frac{2 \cdot \pi \cdot r^3}{3}$$

Remember to use *cubic units* when measuring volume.

❷ Find the volume of each sphere. Use 3.14 for π. Round your answers to the nearest tenth.

(a)
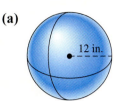

(b)
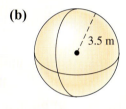

(c) Radius 2.7 cm

ANSWERS
2. (a) $V \approx 7234.6 \text{ in.}^3$
   (b) $V \approx 179.5 \text{ m}^3$
   (c) $V \approx 82.4 \text{ cm}^3$

❸ Find the volume of each hemisphere. Use 3.14 for π. Round your answers to the nearest tenth.

(a)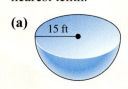

(b)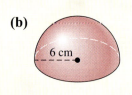

❹ Find the volume of each cylinder. Use 3.14 for π. Round your answers to the nearest tenth.

(a)

(b)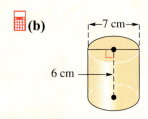

Example 3 Finding the Volume of a Hemisphere

Find the volume of the hemisphere with the help of a calculator. Use 3.14 for π. Round your answer to the nearest tenth.

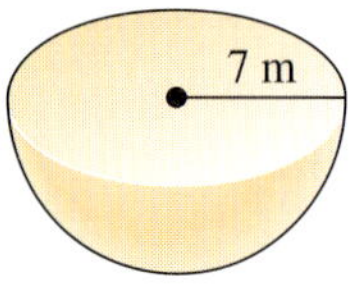

$$V = \frac{2 \cdot \pi \cdot r^3}{3}$$

$$V \approx \frac{2 \cdot 3.14 \cdot 7 \text{ m} \cdot 7 \text{ m} \cdot 7 \text{ m}}{3}$$

$V \approx 718.0 \text{ m}^3$ Rounded to nearest tenth

Work Problem ❸ at the Side.

3 Find the volume of a cylinder. Several *cylinders* are shown here.

These are called *right circular cylinders* because the top and bottom are circles, and the side makes a right angle with the top and bottom. Examples of cylinders are a soup can, a home water heater, and a piece of pipe.

Use the following formula to find the *volume* of a cylinder. Notice that the first part of the formula, $\pi \cdot r \cdot r$, is the area of the circular base.

Finding the Volume of a Cylinder

Volume of cylinder = $\pi \cdot r \cdot r \cdot h$

$V = \pi \cdot r^2 \cdot h$

Remember to use **cubic units** when measuring volume.

Example 4 Finding the Volume of Cylinders

Find the volume of each cylinder. Use 3.14 as the approximate value of π. Round your answers to the nearest tenth if necessary.

(a)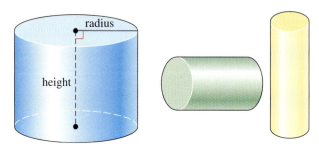

The diameter is 20 m so the radius is $\frac{20 \text{ m}}{2} = 10$ m. The height is 9 m. Use the formula to find the volume.

$V = \pi \cdot r \cdot r \cdot h$
$V \approx 3.14 \cdot 10 \text{ m} \cdot 10 \text{ m} \cdot 9 \text{ m}$
$V \approx 2826 \text{ m}^3$ Cubic units for volume

(b)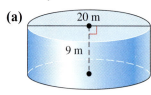

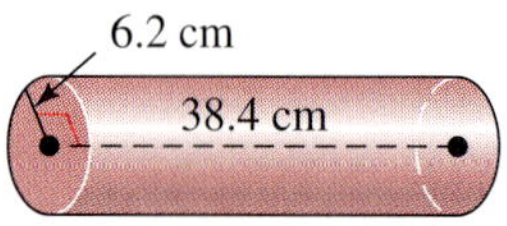

$V \approx 3.14 \cdot 6.2 \text{ cm} \cdot 6.2 \text{ cm} \cdot 38.4 \text{ cm}$
$V \approx 4634.94144$ Now round to tenths.
$V \approx 4634.9 \text{ cm}^3$ Cubic units for volume

Work Problem ❹ at the Side.

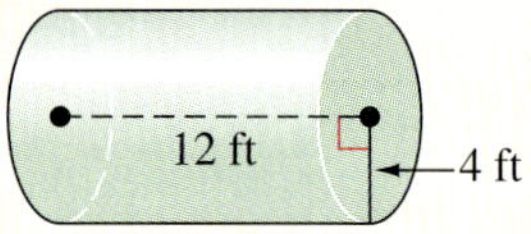

ANSWERS
3. (a) $V \approx 7065 \text{ ft}^3$ (b) $V \approx 452.2 \text{ cm}^3$
4. (a) $V \approx 602.9 \text{ ft}^3$ (b) $V \approx 230.8 \text{ cm}^3$

4 Find the volume of a cone and a pyramid.

A cone and a pyramid are shown here. Notice that the height line is perpendicular to the base in each figure.

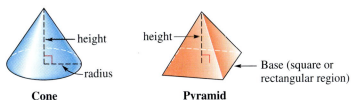

Cone **Pyramid**

Use the following formula to find the *volume* of a cone.

Finding the Volume of a Cone

$$\text{Volume of cone} = \frac{1}{3} \cdot B \cdot h$$

$$\text{or} \quad V = \frac{B \cdot h}{3}$$

where B is the area of the circular base of the cone and h is the height of the cone. Remember to use **cubic units** when measuring volume.

Example 5 **Finding the Volume of a Cone**

Find the volume of the cone. Use 3.14 for π. Round your answer to the nearest tenth.

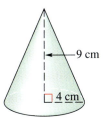

First find the value of B in the formula, which is the *area of the circular base*. Recall that the formula for the area of a circle is πr^2.

$B = \pi \cdot r \cdot r$
$B \approx 3.14 \cdot 4 \text{ cm} \cdot 4 \text{ cm}$
$B \approx 50.24 \text{ cm}^2$ Do not round to tenths yet.

Next, find the volume. The height is 9 cm.

$V = \dfrac{B \cdot h}{3}$

$V \approx \dfrac{50.24 \text{ cm}^2 \cdot 9 \text{ cm}}{3}$

$V \approx 150.72 \text{ cm}^3$ Now round to tenths.
$V \approx 150.7 \text{ cm}^3$ Cubic units for volume

Work Problem 5 at the Side.

---

5 Find the volume of a cone with base radius 2 ft and height 11 ft. Use 3.14 for π. Round your answer to the nearest tenth.

ANSWERS
5. $V \approx 46.1 \text{ ft}^3$

6 Find the volume of a pyramid with base 10 m by 10 m and height 8 m. Round your answer to the nearest tenth.

Use the same formula to find the *volume* of a *pyramid* as you did to find the *volume* of a *cone*.

Finding the Volume of a Pyramid

$$\text{Volume of pyramid} = \frac{1}{3} \cdot B \cdot h$$

$$\text{or} \quad V = \frac{B \cdot h}{3}$$

where B is the area of the square or rectangular base of the pyramid and h is the height of the pyramid. Remember to use **cubic units** when measuring volume.

Example 6 Finding the Volume of a Pyramid

Find the volume of the pyramid. Round your answer to the nearest tenth.

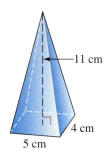

First find the value of B in the formula, which is the *area of the rectangular base*. Recall that the area of a rectangle is found by multiplying length times width.

$$B = 5 \text{ cm} \cdot 4 \text{ cm}$$
$$\mathbf{B = 20 \text{ cm}^2}$$

Next, find the volume.

$$V = \frac{B \cdot h}{3}$$

$$V = \frac{20 \text{ cm}^2 \cdot 11 \text{ cm}}{3}$$

$$V \approx 73.3 \text{ cm}^3 \quad \text{Rounded to nearest tenth}$$

Work Problem **6** **at the Side.**

ANSWERS

6. $V \approx 266.7 \text{ m}^3$

Section 8.7 577

8.7 EXERCISES

FOR EXTRA HELP: Student's Solutions Manual | MyMathLab.com | InterAct Math Tutorial Software | AW Math Tutor Center | www.mathxl.com | Digital Video Tutor CD 5 Videotape 15

Find the volume of each figure. Use 3.14 as the approximate value of π. Round your answers to the nearest tenth if necessary. See Examples 1–6.

1.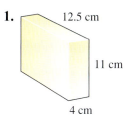

2.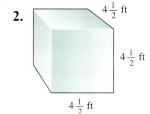

3.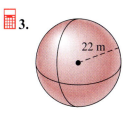

4.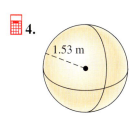

5.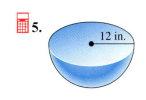

6.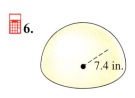

7.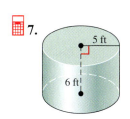

8.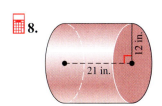

9.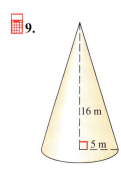

10.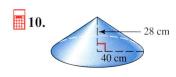

11.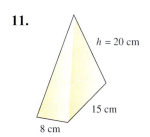

12.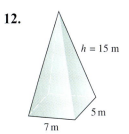

Solve each application problem. Use 3.14 as the approximate value of π. Round your final answers to the nearest tenth if necessary.

13. A pencil box measures 3 in. by 8 in. by $\frac{3}{4}$ in. high. Find the volume of the box. (*Source:* Faber Castell.)

14. A train is being loaded with shipping crates. Each one is 6 m long, 3.4 m wide, and 2 m high. How much space will each crate take?

578 Chapter 8 Geometry

15. An oil candle globe made of hand-blown glass has a diameter of 16.8 cm. What is the volume of the globe?

16. A metal sphere used as part of a fountain has a diameter of $6\frac{1}{2}$ ft. Find its volume.

17. One of the ancient stone pyramids in Egypt has a square base that measures 145 m on each side. The height is 93 m. What is the volume of the pyramid? (*Source: The Columbia Encyclopedia.*)

18. A cylindrical woven basket made by a Northwest Coast tribe is 8 cm high and has a diameter of 11 cm. What is the volume of the basket?

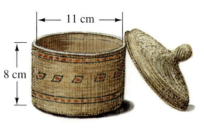

19. A city sewer pipe has a diameter of 5 ft and a length of 200 ft. Find the volume of the pipe.

20. An ice cream cone has a diameter of 2 in. and a height of 4 in. Find its volume.

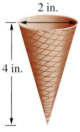

21. Explain the *two* errors made by a student in finding the volume of a cylinder with a diameter of 7 cm and a height of 5 cm. Find the correct answer.

$$V \approx 3.14 \cdot 7 \cdot 7 \cdot 5$$
$$V \approx 769.3 \text{ cm}^2$$

22. Compare the steps in finding the volume of a cylinder and a cone. How are they similar? Suppose you know the volume of a cylinder. How can you find the volume of a cone with the same radius and height by doing just a one-step calculation?

23. Find the volume.

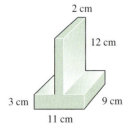

24. Find the volume. (*Hint:* Notice the hole that goes through the center of the shape.)

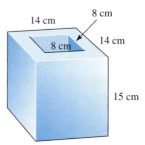

8.8 PYTHAGOREAN THEOREM

OBJECTIVES
1. Find square roots using the square root key on a calculator.
2. Find the unknown length in a right triangle.
3. Solve application problems involving right triangles.

In **Section 8.3** you used this formula for the area of a square, $A = s^2$. The square below on the left has an area of 25 cm² because 5 cm • 5 cm = 25 cm².

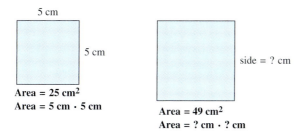

The square on the right has an area of 49 cm². To find the length of a side, ask yourself, "What number can be multiplied by itself to give 49?" Because 7 • 7 = 49, the length of each side is 7 cm.

Remember: 7 • 7 = 49, so 7 is the **square root** of 49, or $\sqrt{49} = 7$. Also, $\sqrt{81} = 9$, since 9 • 9 = 81. (See **Section 1.8** for further review.)

Work Problem ❶ at the Side.

A number that has a whole number as its square root is called a *perfect square*. For example, 9 is a perfect square because $\sqrt{9} = 3$, and 3 is a whole number.

The first few perfect squares are listed here.

Some Perfect Squares

$\sqrt{1} = 1$	$\sqrt{16} = 4$	$\sqrt{49} = 7$	$\sqrt{100} = 10$
$\sqrt{4} = 2$	$\sqrt{25} = 5$	$\sqrt{64} = 8$	$\sqrt{121} = 11$
$\sqrt{9} = 3$	$\sqrt{36} = 6$	$\sqrt{81} = 9$	$\sqrt{144} = 12$

1 **Find square roots using the square root key on a calculator.** If a number is *not* a perfect square, then you can find its *approximate* square root by using a calculator with a square root key.

Calculator Tip To find a square root, use the √ key on a standard calculator or the √x key on a scientific calculator. In either case, you do *not* need to use the = key. Try these. Jot down your answers.

To find $\sqrt{16}$ press: 16 √x   Answer is 4.

To find $\sqrt{7}$ press: 7 √x   Answer is 2.645751311.

For $\sqrt{7}$, your calculator shows 2.645751311, which is an *approximate* answer. We will be rounding to the nearest thousandth, so $\sqrt{7} \approx 2.646$. To check, multiply 2.646 times 2.646. Do you get 7 as the result? No, you get 7.001316, which is very close to 7. The difference is due to rounding.

❶ Find each square root.

(a) $\sqrt{36}$

(b) $\sqrt{25}$

(c) $\sqrt{9}$

(d) $\sqrt{100}$

(e) $\sqrt{121}$

ANSWERS
1. (a) 6 (b) 5 (c) 3 (d) 10 (e) 11

2 Use a calculator with a square root key to find each square root. Round to the nearest thousandth if necessary.

(a) $\sqrt{11}$

(b) $\sqrt{40}$

(c) $\sqrt{56}$

(d) $\sqrt{196}$

(e) $\sqrt{147}$

Example 1 Finding the Square Root of Numbers

Use a calculator to find each square root. Round your answers to the nearest thousandth.

(a) $\sqrt{35}$ Calculator shows 5.916079783; round to 5.916

(b) $\sqrt{124}$ Calculator shows 11.13552873; round to 11.136

(c) $\sqrt{200}$ Calculator shows 14.14213562; round to 14.142

Work Problem **2** at the Side.

2 **Find the unknown length in a right triangle.** One place you will use square roots is when working with the *Pythagorean Theorem*. This theorem applies only to *right* triangles (triangles with a 90° angle). The longest side of a right triangle is called the **hypotenuse.** It is opposite the right angle. The other two sides are called *legs*. The legs form the right angle.

Examples of right triangles

Pythagorean Theorem

$$(\text{hypotenuse})^2 = (\text{leg})^2 + (\text{leg})^2$$

In other words, square the length of each side. After you have squared all the sides, the sum of the squares of the two legs will equal the square of the hypotenuse.

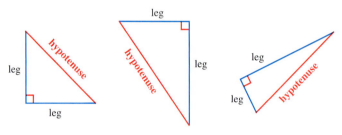

$(\text{hypotenuse})^2 = (\text{leg})^2 + (\text{leg})^2$

$5^2 = 4^2 + 3^2$

$25 = 16 + 9$

$25 = 25$

The theorem is named after Pythagoras, a Greek mathematician who lived about 2500 years ago. He and his followers may have used floor tiles to prove the theorem, as shown below.

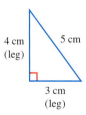
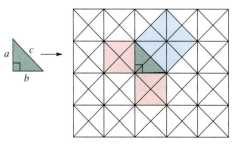

The right triangle in the center of the floor tiles has sides a, b, and c. The square drawn on side a contains four triangular tiles. The square on side b contains four tiles. The square on side c contains eight tiles. The number of tiles in the square on side c equals the sum of the number of tiles in the squares on sides a and b, that is, 8 tiles = 4 tiles + 4 tiles. As a result, you often see the Pythagorean Theorem written as $c^2 = a^2 + b^2$.

ANSWERS

2. (a) $\sqrt{11} \approx 3.317$ (b) $\sqrt{40} \approx 6.325$
   (c) $\sqrt{56} \approx 7.483$ (d) $\sqrt{196} = 14$
   (e) $\sqrt{147} \approx 12.124$

If you know the lengths of any two sides in a right triangle, you can use the Pythagorean Theorem to find the length of the third side.

Using the Pythagorean Theorem

To find the hypotenuse, use this formula:

$$\text{hypotenuse} = \sqrt{(\text{leg})^2 + (\text{leg})^2}$$

To find a leg, use this formula:

$$\text{leg} = \sqrt{(\text{hypotenuse})^2 - (\text{leg})^2}$$

CAUTION

Remember: A small square drawn in one angle of a triangle indicates a right angle. You can use the Pythagorean Theorem *only* on triangles that have a right angle.

Example 2 Finding the Unknown Length in Right Triangles

Find the unknown length in each right triangle. Round answers to the nearest tenth if necessary.

(a) 3 ft, 4 ft

The length of the side opposite the right angle is unknown. That side is the hypotenuse, so use this formula.

$$\begin{aligned}
\text{hypotenuse} &= \sqrt{(\text{leg})^2 + (\text{leg})^2} & &\text{Find the hypotenuse.}\\
\text{hypotenuse} &= \sqrt{(3)^2 + (4)^2} & &\text{Legs are 3 and 4.}\\
&= \sqrt{9 + 16} & &\text{3 • 3 is 9 and 4 • 4 is 16.}\\
&= \sqrt{25}\\
&= 5
\end{aligned}$$

The hypotenuse is 5 ft long.

(b) 15 cm, 7 cm

We *do* know the length of the hypotenuse (15 cm), so it is the length of one of the legs that is unknown. Use this formula.

$$\begin{aligned}
\text{leg} &= \sqrt{(\text{hypotenuse})^2 - (\text{leg})^2} & &\text{Find a leg.}\\
\text{leg} &= \sqrt{(15)^2 - (7)^2} & &\text{Hypotenuse is 15, one leg is 7.}\\
&= \sqrt{225 - 49} & &\text{15 • 15 is 225 and 7 • 7 is 49.}\\
&= \sqrt{176} & &\text{Use calculator to find }\sqrt{176}.\\
&\approx 13.3 & &\text{Round 13.26649916 to 13.3.}
\end{aligned}$$

The length of the leg is approximately 13.3 cm.

→ Work Problem ❸ at the Side.

CAUTION

You use the Pythagorean Theorem to find the *length* of one side, *not* the area of the triangle. Your answer will be in linear units, such as ft, yd, cm, m, and so on (*not* ft^2, yd^2, cm^2, m^2).

❸ Find the unknown length in each right triangle. Round your answers to the nearest tenth if necessary.

(a) 5 in., 12 in.

(b)  7 cm, 25 cm, 90°

(c) 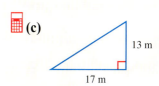 13 m, 17 m

(d) 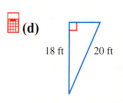 18 ft, 20 ft

(e) 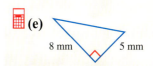 8 mm, 5 mm

Answers

3. **(a)** $\sqrt{169} = 13$ in.   **(b)** $\sqrt{576} = 24$ cm
   (c) $\sqrt{458} \approx 21.4$ m   **(d)** $\sqrt{76} \approx 8.7$ ft
   (e) $\sqrt{89} \approx 9.4$ mm

4 These problems show ladders leaning against buildings. Find the unknown lengths. Round to the nearest tenth of a foot if necessary.

(a)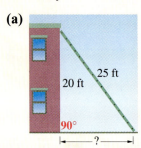

How far away from the building is the bottom of the ladder?

(b)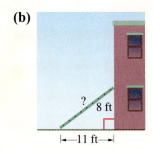

How long is the ladder?

(c) A 17-ft ladder is leaning against a building. The bottom of the ladder is 10 ft from the building. How high up on the building will the ladder reach? (*Hint:* Start by drawing the building and the ladder.)

3 Solve application problems involving right triangles. The next example shows an application of the Pythagorean Theorem.

Example 3   Using the Pythagorean Theorem

A television antenna is on the roof of a house, as shown. Find the length of the support wire. Round your answer to the nearest tenth of a meter if necessary.

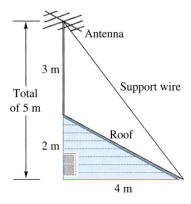

A right triangle is formed. The total length of the side at the left is 3 m + 2 m = 5 m.

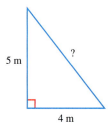

The support wire is the hypotenuse of the right triangle.

$$\text{hypotenuse} = \sqrt{(\text{leg})^2 + (\text{leg})^2} \quad \text{Find the hypotenuse.}$$
$$\text{hypotenuse} = \sqrt{(5)^2 + (4)^2} \quad \text{Legs are 5 and 4.}$$
$$= \sqrt{25 + 16} \quad 5^2 \text{ is 25 and } 4^2 \text{ is 16.}$$
$$= \sqrt{41} \quad \text{Use } \sqrt{x} \text{ key on a calculator.}$$
$$\approx 6.4 \quad \text{Round 6.403124237 to 6.4.}$$

The length of the support wire is approximately 6.4 m.

Work Problem 4 at the Side.

ANSWERS
4. (a) $\sqrt{225} = 15$ ft (b) $\sqrt{185} \approx 13.6$ ft
   (c) $\sqrt{189} \approx 13.7$ ft

8.8 EXERCISES

FOR EXTRA HELP: Student's Solutions Manual | MyMathLab.com | InterAct Math Tutorial Software | AW Math Tutor Center | www.mathxl.com | Digital Video Tutor CD 5 Videotape 15

Find each square root. Starting with Exercise 5, use the square root key on a calculator. Round your answers to the nearest thousandth if necessary. See Example 1.

1. $\sqrt{16}$ **2.** $\sqrt{4}$ **3.** $\sqrt{64}$ **4.** $\sqrt{81}$

5. $\sqrt{11}$ **6.** $\sqrt{23}$ **7.** $\sqrt{5}$ **8.** $\sqrt{2}$

9. $\sqrt{73}$ **10.** $\sqrt{80}$ **11.** $\sqrt{101}$ **12.** $\sqrt{125}$

13. $\sqrt{190}$ **14.** $\sqrt{160}$ **15.** $\sqrt{1000}$ **16.** $\sqrt{2000}$

17. You know that $\sqrt{25} = 5$ and $\sqrt{36} = 6$. Using just that information (no calculator), describe how you could estimate $\sqrt{30}$. How would you estimate $\sqrt{26}$ or $\sqrt{35}$? Now check your estimates using a calculator.

18. Explain the relationship between *squaring* a number and finding the *square root* of a number. Include two examples to illustrate your explanation.

Find the unknown length in each right triangle. Use a calculator to find square roots. Round your answers to the nearest tenth if necessary. See Example 2.

19.

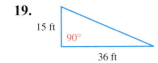

20.

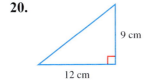

21.

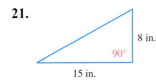

22.

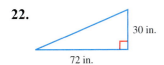

23.

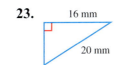

24.

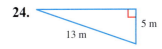

584 Chapter 8 Geometry

25. 3 in.

8 in.

26.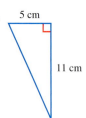

5 cm

11 cm

27.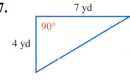

7 yd

90°

4 yd

28.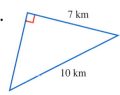

7 km

10 km

29.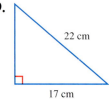

22 cm

17 cm

30.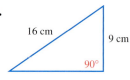

16 cm

9 cm

90°

31.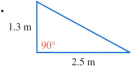

1.3 m

90°

2.5 m

32.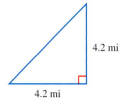

4.2 mi

4.2 mi

33.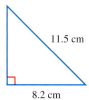

11.5 cm

8.2 cm

34.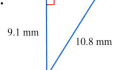

9.1 mm

10.8 mm

35.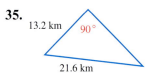

13.2 km 90°

21.6 km

36.

26.5 ft

37.4 ft

Solve each application problem. Round your answers to the nearest tenth if necessary. See Example 3.

37. Find the length of this loading ramp.

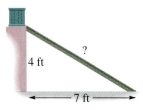

4 ft

?

7 ft

38. Find the unknown length in this roof plan.

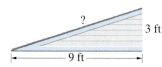

?

3 ft

9 ft

39. How high is the airplane above the ground?

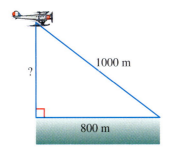

40. Find the height of this farm silo.

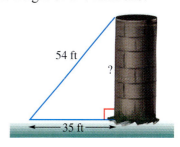

41. To reach his lady-love, a knight placed a 12-ft ladder against the castle wall. If the base of the ladder is 3 ft from the building, how high on the castle will the top of the ladder reach? Draw a sketch of the castle and ladder and solve the problem.

42. William drove his car 15 miles north, then made a right turn and drove 7 miles east. How far is he, in a straight line, from his starting point? Draw a sketch to illustrate the problem and solve it.

43. Describe the *two* errors made by a student in solving this problem. Also find the correct answer. Round to the nearest tenth.

$$? = \sqrt{(9)^2 + (7)^2}$$
$$= \sqrt{18 + 14}$$
$$= \sqrt{32} \approx 5.657 \text{ in.}$$

9 in.
?
7 in.

44. Describe the *two* errors made by a student in solving this problem. Also find the correct answer. Round to the nearest tenth.

$$? = \sqrt{(13)^2 + (20)^2}$$
$$= \sqrt{169 + 400}$$
$$= \sqrt{569} \approx 23.9 \text{ m}^2$$

RELATING CONCEPTS (Exercises 45–48) FOR INDIVIDUAL OR GROUP WORK

*Use your knowledge of the Pythagorean Theorem to **work Exercises 45–48 in order.** Round answers to the nearest tenth.*

45. A major league baseball diamond is a square shape measuring 90 ft on each side. If the catcher throws a ball from home plate to second base, how far is he throwing the ball? (*Source:* American League of Professional Baseball Clubs.)

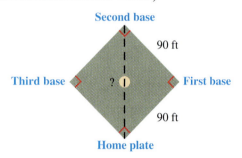

46. A softball diamond is only 60 ft on each side. (*Source:* Amateur Softball Association.)

(a) Draw a sketch of the diamond and label the sides and bases.

(b) How far is it to throw a ball from home plate to second base?

47. Look back at your answer to Exercise 45. Explain how you can tell the distance from third base to first base without doing any further calculations.

48. Show how you could set up a proportion to answer Exercise 46 instead of using the Pythagorean Theorem. (You'll need your answer from Exercise 45.)

8.9 SIMILAR TRIANGLES

Two triangles with the same *shape* (but not necessarily the same size) are called **similar triangles.** Three pairs of similar triangles are shown below.

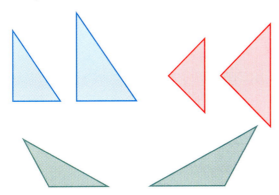

OBJECTIVES

1. Identify corresponding parts in similar triangles.
2. Find the unknown lengths of sides in similar triangles.
3. Solve application problems involving similar triangles.

1 Identify corresponding parts in similar triangles. The two triangles shown below are different sizes but have the same shape, so they are *similar triangles*. Angles *A* and *P* measure the same number of degrees and are called *corresponding angles*. Angles *B* and *Q* are corresponding angles, as are angles *C* and *R*. The triangles have the same shape because the corresponding angles have the same measure.

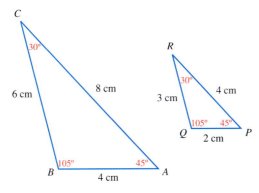

$\overline{PR}$ and $\overline{AC}$ are called *corresponding sides* because they are *opposite* corresponding angles. Also, $\overline{QR}$ and $\overline{BC}$ are corresponding sides, as are $\overline{PQ}$ and $\overline{AB}$. Although corresponding angles measure the same number of degrees, corresponding sides do *not* need to be the same length. In the triangles here, each side in the smaller triangle is *half* the length of the corresponding side in the larger triangle.

Work Problem ❶ at the Side.

2 Find the unknown lengths of sides in similar triangles. Similar triangles are useful because of the following property.

Similar Triangles

In **similar triangles,** the ratios of the lengths of corresponding sides are equal.

❶ Identify corresponding angles and sides in these similar triangles.

(a)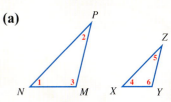

Angles:
1 and _____
2 and _____
3 and _____

Sides:
$\overline{PN}$ and _____
$\overline{PM}$ and _____
$\overline{NM}$ and _____

(b)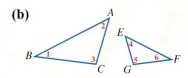

Angles:
1 and _____
2 and _____
3 and _____

Sides:
$\overline{AB}$ and _____
$\overline{BC}$ and _____
$\overline{AC}$ and _____

ANSWERS
1. (a) 4; 5; 6; $\overline{ZX}$; $\overline{ZY}$; $\overline{XY}$
   (b) 6; 4; 5; $\overline{EF}$; $\overline{FG}$; $\overline{EG}$

② Find the length of x in Example 1 by setting up and solving a proportion.

Example 1 — Finding the Unknown Lengths of Sides in Similar Triangles

Find the length of y in the smaller triangle. Assume the triangles are similar.

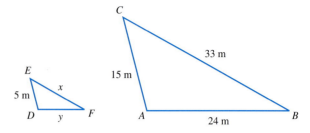

$\overline{ED}$ and $\overline{CA}$ are corresponding sides. The ratio of the lengths of these sides can be written as a fraction in lowest terms.

$$\frac{ED}{CA} = \frac{5 \text{ m}}{15 \text{ m}} = \frac{1}{3} \leftarrow \text{Lowest terms}$$

As mentioned earlier, the ratios of the lengths of corresponding sides are equal. $\overline{DF}$ in the smaller triangle corresponds to $\overline{AB}$ in the larger triangle. Since the ratios of corresponding sides are equal,

$$\frac{DF}{AB} = \frac{1}{3}$$

Replace DF with y and AB with 24 to get this proportion.

$$\frac{y}{24} = \frac{1}{3}$$

Find the cross products.

$$\frac{y}{24} = \frac{1}{3} \qquad \begin{array}{l} 24 \cdot 1 = 24 \\ y \cdot 3 \end{array}$$

Show that the cross products are equivalent.

$$y \cdot 3 = 24$$

Divide each side by 3.

$$\frac{y \cdot \cancel{3}}{\cancel{3}} = \frac{24}{3}$$

$$y = 8$$

$\overline{DF}$ has a length of 8 m.

Work Problem ② at the Side.

ANSWERS

2. $\dfrac{x}{33} = \dfrac{1}{3}$; $x = 11$ m

Example 2 Finding an Unknown Length and the Perimeter

Find the perimeter of the smaller triangle. Assume the triangles are similar.

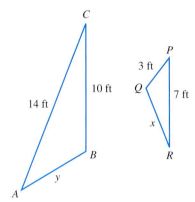

First find the length of $\overline{RQ}$, then add up the sides to find the perimeter. The smaller triangle is turned "upside down" compared to the larger triangle, so be careful when identifying corresponding sides. $\overline{AC}$ is the longest side in the larger triangle, and $\overline{PR}$ is the longest side in the smaller triangle. So $\overline{PR}$ and $\overline{AC}$ are corresponding sides. The ratio of their lengths can be written as a fraction in lowest terms.

$$\frac{PR}{AC} = \frac{7 \text{ ft}}{14 \text{ ft}} = \frac{1}{2} \quad \leftarrow \text{Lowest terms}$$

The two triangles are similar, so the ratio of any pair of corresponding sides will also equal $\frac{1}{2}$. Because $\overline{RQ}$ and $\overline{CB}$ are corresponding sides,

$$\frac{RQ}{CB} = \frac{1}{2}.$$

Replace RQ with x and CB with 10 to make a proportion.

$$\frac{x}{10} = \frac{1}{2}$$

Find the cross products.

$$10 \cdot 1 = 10$$
$$\frac{x}{10} = \frac{1}{2}$$
$$x \cdot 2$$

Show that the cross products are equivalent.

$$x \cdot 2 = 10$$

Divide each side by 2.

$$\frac{x \cdot \cancel{2}}{\cancel{2}} = \frac{10}{2}$$

$$x = 5$$

$\overline{RQ}$ has a length of 5 ft. Now add the lengths of all three sides to find the perimeter.

$$\text{perimeter} = 5 \text{ ft} + 3 \text{ ft} + 7 \text{ ft} = 15 \text{ ft}$$

Work Problem 3 at the Side.

3 (a) Find the perimeter of triangle ABC in Example 2.

(b) Find the perimeter of each triangle. Assume the triangles are similar.

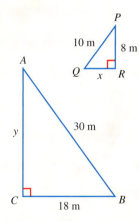

ANSWERS

3. (a) $y = 6$ ft; perimeter = 14 ft + 10 ft + 6 ft = 30 ft
   (b) $x = 6$ m, perimeter = 24 m; $y = 24$ m, perimeter = 72 m

4 Find the height of each flagpole.

(a)
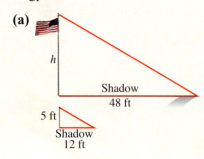

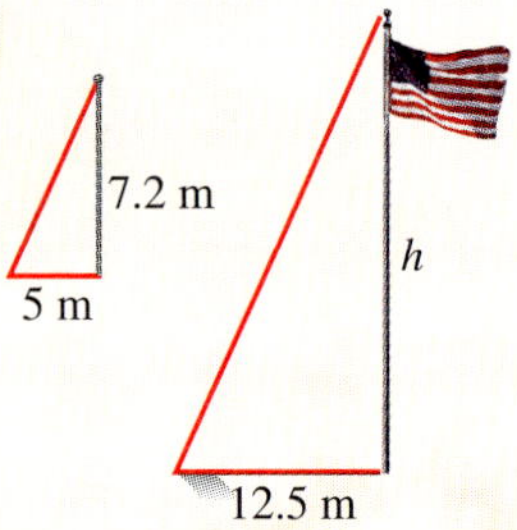

3 Solve application problems involving similar triangles. The next example shows an application of similar triangles.

Example 3 Using Similar Triangles in an Application

A flagpole casts a shadow 99 m long at the same time that a pole 10 m tall casts a shadow 18 m long. Find the height of the flagpole.

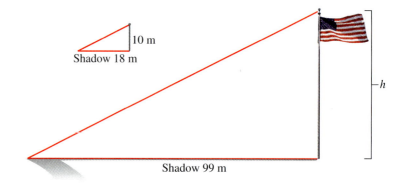

The triangles shown are similar, so write a proportion to find h.

Height in larger triangle → $\dfrac{h}{10} = \dfrac{99}{18}$ ← Shadow in larger triangle
Height in smaller triangle →   ← Shadow in smaller triangle

Find the cross products and show that they are equivalent.

$$h \cdot 18 = 10 \cdot 99$$
$$h \cdot 18 = 990$$

Divide each side by 18.

$$\dfrac{h \cdot \cancel{18}}{\cancel{18}} = \dfrac{990}{18}$$

$$h = 55$$

The flagpole is 55 m high.

NOTE

There are several other correct ways to set up the proportion in Example 3. One way is to simply flip the ratios on *both* sides of the equal sign.

$$\dfrac{10}{h} = \dfrac{18}{99}$$

But there is another option, shown below.

Height in larger triangle → $\dfrac{h}{99} = \dfrac{10}{18}$ ← Height in smaller triangle
Shadow in larger triangle →   ← Shadow in smaller triangle

Notice that both ratios compare *height* to *shadow* in the same order. The ratio on the left describes the larger triangle, and the ratio on the right describes the smaller triangle.

Work Problem 4 **at the Side.**

(b)

7.2 m
5 m
h
12.5 m

ANSWERS
4. (a) $h = 20$ ft (b) $h = 18$ m

Section 8.9 **591**

8.9 EXERCISES

FOR EXTRA HELP: Student's Solutions Manual • MyMathLab.com • InterAct Math Tutorial Software • AW Math Tutor Center • www.mathxl.com • Digital Video Tutor CD 5 Videotape 15

Write similar *or* not similar *for each pair of triangles.*

1.

2.

3.

4.

5.

6.

Name the corresponding angles and the corresponding sides in each pair of similar triangles. See Margin Problem 1.

7.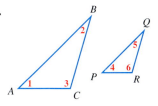

8.

9.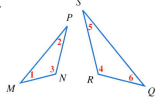

10.

Write the ratio for each pair of corresponding sides in the similar triangles shown below. Write the ratios as fractions in lowest terms. See Examples 1 and 2.

11. $\dfrac{AB}{PQ}$; $\dfrac{AC}{PR}$; $\dfrac{BC}{QR}$

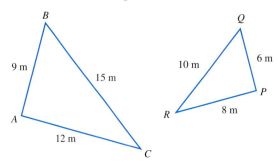

12. $\dfrac{AB}{PQ}$; $\dfrac{AC}{PR}$; $\dfrac{BC}{QR}$

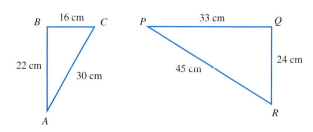

Find the unknown lengths in each pair of similar triangles. See Example 1.

13.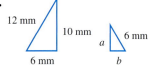

14.

15.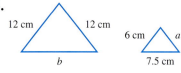

16.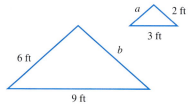

Find the perimeter of each triangle. Assume the triangles are similar. See Example 2.

17.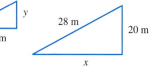

18.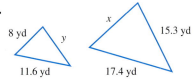

19. Triangles *CDE* and *FGH* are similar. Find the perimeter and area of triangle *FGH*. *Note:* The heights of similar triangles have the same ratio as corresponding sides.

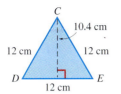

20. Triangles *JKL* and *MNO* are similar. Find the perimeter and area of triangle *MNO*.

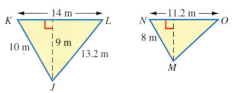

Solve each application problem. See Example 3.

21. The height of the house shown here can be found by comparing its shadow to the shadow cast by a 3 ft stick. Find the height of the house by writing a proportion and solving it.

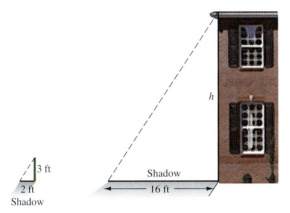

22. A fire lookout tower provides an excellent view of the surrounding countryside. The height of the tower can be found by lining up the top of the tower with the top of a 2 m stick. Use similar triangles to find the height of the tower.

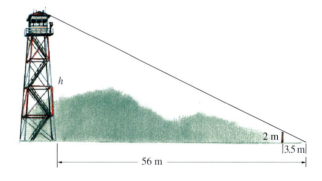

23. Refer to the building in Exercise 21. Later in the day, the same building had a shadow 6 ft long. How long would the stick's shadow be at that time?

24. Refer to the lookout tower in Exercise 22.

(a) How far away from the tower was the 2 m stick?

(b) To use a 1 m stick and have it line up with the same endpoint, how far from the tower would it have to be?

25. Look up the word *similar* in a dictionary. What is the nonmathematical definition of this word? Describe two examples of similar objects at home, school, or work.

26. *Congruent* objects have the *same shape* and the *same size*. Sketch a pair of congruent triangles. Describe two examples of congruent objects at home, school, or work.

Find the unknown length in Exercises 27–30. Round your answers to the nearest tenth. Note: When a line is drawn parallel to one side of a triangle, the smaller triangle that is formed will be similar to the original triangle. In Exercises 27–28, the red segments are parallel.

27.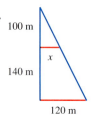

28.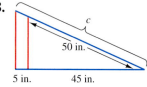

29. Use similar triangles and a proportion to find the length of the lake shown here. (*Hint:* The side 100 m long in the smaller triangle corresponds to a side of 100 m + 120 m = 220 m in the larger triangle.)

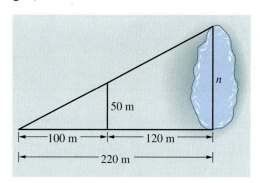

30. To find the height of the tree, find y and then add $5\frac{1}{2}$ ft for the distance from the ground to the person's eye level.

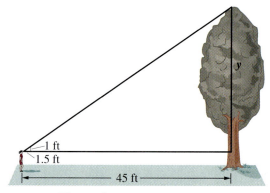

Distance from person to tree.

Chapter 8

SUMMARY

KEY TERMS

8.1

point	A point is a location in space. *Example:* Point *P* at the right.	
line	A line is a straight row of points that goes on forever in both directions. *Example:* Line *AB*, written $\overleftrightarrow{AB}$, at the right.	
line segment	A line segment is a piece of a line with two endpoints. *Example:* Line segment *PQ*, written $\overline{PQ}$, at the right.	
ray	A ray is a part of a line that has one endpoint and extends forever in one direction. *Example:* Ray *RS*, written $\overrightarrow{RS}$, at the right.	
angle	An angle is made up of two rays that have a common endpoint called the vertex. *Example:* Angle 1 at the right.	
degrees	A system used to measure angles in which a complete circle is 360 degrees, written 360°.	
right angle	A right angle is an angle that measures exactly 90°. *Example:* Angle *AOB* at right.	
acute angle	An acute angle is an angle that measures less than 90°. *Example:* Angle *E* at the right.	
obtuse angle	An obtuse angle is an angle that measures more than 90° but less than 180°. *Example:* Angle *F* at the right.	
straight angle	A straight angle is an angle that measures 180°; its sides form a straight line. *Example:* Angle *G* at the right.	
intersecting lines	Intersecting lines cross or merge. *Example:* $\overleftrightarrow{RQ}$ intersects $\overleftrightarrow{AB}$ at point *P* at the right.	
perpendicular lines	Perpendicular lines are two lines that intersect to form a right angle. *Example:* $\overleftrightarrow{PQ}$ is perpendicular to $\overleftrightarrow{RS}$ at the right.	
parallel lines	Parallel lines are two lines in the same plane that never intersect (never cross). *Example:* $\overleftrightarrow{AB}$ is parallel to $\overleftrightarrow{ST}$ at the right.	

8.2

complementary angles	Complementary angles are two angles whose measures add up to 90°.
supplementary angles	Supplementary angles are two angles whose measures add up to 180°.
congruent angles	Congruent angles are angles that measure the same number of degrees.

Key Terms

	vertical angles	Vertical angles are two nonadjacent congruent angles formed by intersecting lines. *Example:* $\angle COA$ and $\angle EOF$ are vertical angles at the right.
8.3–8.5	perimeter	Perimeter is the distance around the outside edges of a flat shape. It is measured in linear units such as ft, yd, cm, m, km, and so on.
8.3–8.6	area	Area is the surface inside a two-dimensional (flat) shape. It is measured by determining the number of squares of a certain size needed to cover the surface inside the shape. Some of the commonly used units for measuring area are square inches (in.^{2}), square feet (ft^2), square yards (yd^2), square centimeters (cm^2), and square meters (m^2).
8.3	rectangle	A rectangle is a four-sided figure with all sides meeting at 90° angles. The opposite sides are the same length. *Example:* A rectangle measuring 12 cm by 7 cm at the right.
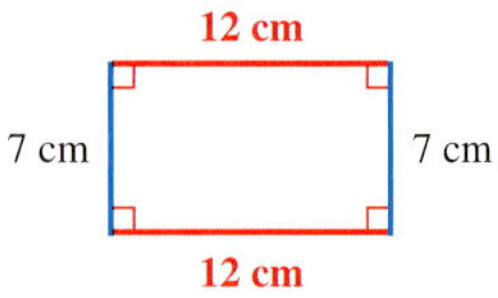	square	A square is a rectangle with all four sides the same length. *Example:* A square with the side measurement of 20 in. at the right.
8.4	parallelogram	A parallelogram is a four-sided figure with both pairs of opposite sides parallel. *Example:* See figure at the right with sides measuring 8 m and 12 m.
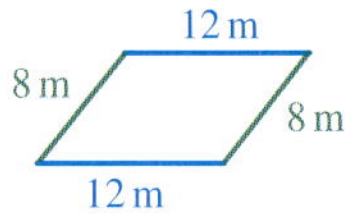	trapezoid	A trapezoid is a four-sided figure with one pair of parallel sides. *Example:* Trapezoid *PQRS* at the right; $\overline{PQ}$ is parallel to $\overline{SR}$.
8.5	triangle	A triangle is a figure with exactly three sides. *Example:* Triangle *ABC* at the right.
8.6	circle	A circle is a figure with all points the same distance from a fixed center point. *Example:* See figure at the right.
	radius	Radius is the distance from the center of a circle to any point on the circle. *Example:* See the red radius line in the circle at the right.
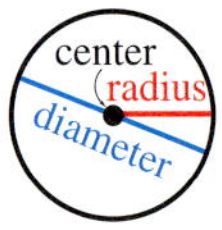	diameter	Diameter is the distance across a circle, passing through the center. *Example:* See the blue diameter line in the circle at the right.
	circumference	Circumference is the distance around a circle.
	π (pi)	π is the ratio of the circumference to the diameter of any circle. It is approximately equal to 3.14.
8.7	volume	Volume is a measure of the space inside a three-dimensional (solid) shape. Volume is measured in cubic units such as in.^{3}, ft^3, yd^3, mm^3, cm^3, and so on.
8.8	square root	A square root is one of two equal factors of a number.
	hypotenuse	The hypotenuse is the side of a right triangle opposite the 90° angle; it is the longest side. *Example:* See figure at the right.
8.9	similar triangles	Similar triangles are triangles with the same shape but not necessarily the same size; corresponding angles measure the same number of degrees, and the *ratios* of the lengths of corresponding sides are equal.

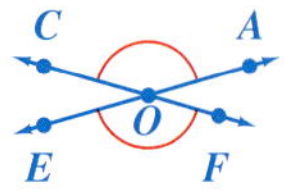

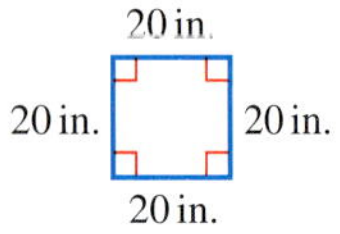

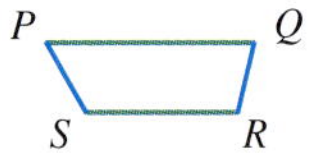

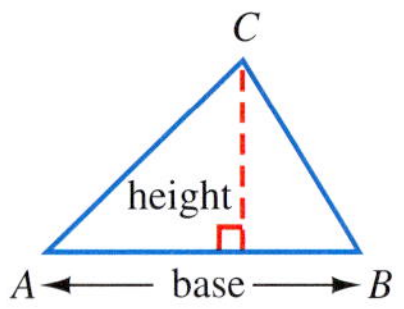

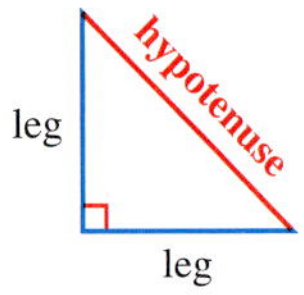

New Formulas

Perimeter of a rectangle: $P = 2 \cdot l + 2 \cdot w$

Area of a rectangle: $A = l \cdot w$

Perimeter of a square: $P = 4 \cdot s$

Area of a square: $A = s^2$

Area of a parallelogram: $A = b \cdot h$

Area of a trapezoid: $A = \frac{1}{2} \cdot h \cdot (b + B)$

or $A = 0.5 \cdot h \cdot (b + B)$

Area of a triangle: $A = \frac{1}{2} \cdot b \cdot h$

or $A = 0.5 \cdot b \cdot h$

Diameter of a circle: $d = 2 \cdot r$

Radius of a circle: $r = \frac{d}{2}$

Circumference of a circle: $C = \pi \cdot d$

or $C = 2 \cdot \pi \cdot r$

Area of a circle: $A = \pi \cdot r^2$

Area of semicircle: $A = \frac{\pi \cdot r^2}{2}$

Volume of rectangular solid: $V = l \cdot w \cdot h$

Volume of a sphere: $V = \frac{4}{3} \cdot \pi \cdot r^3$

or $V = \frac{4 \cdot \pi \cdot r^3}{3}$

Volume of a hemisphere: $V = \frac{2}{3} \cdot \pi \cdot r^3$

or $V = \frac{2 \cdot \pi \cdot r^3}{3}$

Volume of a cylinder: $V = \pi \cdot r^2 \cdot h$

Volume of a cone: $V = \frac{1}{3} \cdot B \cdot h$ or $\frac{B \cdot h}{3}$

Volume of a pyramid: $V = \frac{1}{3} \cdot B \cdot h$ or $\frac{B \cdot h}{3}$

Right triangle: hypotenuse $= \sqrt{(\text{leg})^2 + (\text{leg})^2}$

leg $= \sqrt{(\text{hypotenuse})^2 - (\text{leg})^2}$

Test Your Word Power

See how well you have learned the vocabulary in this chapter. Answers follow the Quick Review.

1. Two angles that are **complementary**
   (a) have measures that add up to 180°
   (b) are always congruent
   (c) form a straight angle
   (d) have measures that add up to 90°.

2. The **perimeter** of a flat shape is
   (a) measured in square units
   (b) the distance around the outside edges
   (c) the number of squares needed to cover the space inside the shape
   (d) measured in cubic units.

3. An **obtuse angle**
   (a) is formed by perpendicular lines
   (b) is congruent to a right angle
   (c) measures more than 90° but less than 180°
   (d) measures less than 90°.

4. The **hypotenuse** is
   (a) the long base in a trapezoid
   (b) the height line in a parallelogram
   (c) the longest side in a right triangle
   (d) the distance across a circle, passing through the center.

5. π is the ratio of
   (a) the diameter to the radius of a circle
   (b) the circumference to the diameter of a circle
   (c) the circumference to the radius of a circle
   (d) the diameter to the circumference of a circle.

6. **Perpendicular lines**
   (a) intersect to form a right angle
   (b) intersect to form an acute angle
   (c) never intersect
   (d) have a common endpoint called the vertex.

7. In a pair of **similar triangles,**
   (a) corresponding sides have the same length
   (b) all the angles have the same measure
   (c) the perimeters are equal
   (d) the ratios of the lengths of corresponding sides are equal.

8. The **area of a rectangle** is found by
   (a) using the formula $P = 2 \cdot l + 2 \cdot w$
   (b) multiplying length times width
   (c) adding the lengths of the sides
   (d) using the formula $V = l \cdot w \cdot h$.

Quick Review

Concepts

8.1 Lines

If a line has one endpoint, it is a *ray*. If it has two endpoints, it is a *line segment*.

If two lines intersect at right angles, they are *perpendicular*.

If two lines in the same plane never intersect, they are *parallel*.

Examples

Identify each of the following as a line, line segment, or ray.

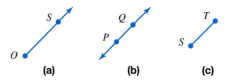

Figure **(a)** shows a ray, **(b)** shows a line, and **(c)** shows a line segment.

Label each pair of lines as parallel or perpendicular.

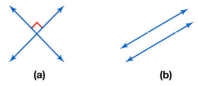

Figure **(a)** shows two perpendicular lines (they intersect at 90°).

Figure **(b)** shows two parallel lines (they never intersect).

8.2 Angles

If the sum of the measures of two angles is 90°, they are *complementary*.

If the sum of the measures of two angles is 180°, they are *supplementary*.

If two angles measure the same number of degrees, the angles are *congruent*. The symbol for congruent is ≅.

Two nonadjacent angles formed by intersecting lines are called *vertical angles*. Vertical angles are congruent.

Find the complement and supplement of a 35° angle.

$$90° - 35° = 55° \text{ (the complement)}$$
$$180° - 35° = 145° \text{ (the supplement)}$$

Identify the vertical angles in this figure. Which angles are congruent?

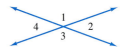

∠1 and ∠3 are vertical angles.

∠2 and ∠4 are vertical angles.

Vertical angles are congruent, so ∠1 ≅ ∠3 and ∠2 ≅ ∠4.

Chapter 8 Summary

Concepts	Examples
8.3 Rectangles and Squares	

Use this formula to find the perimeter of a *rectangle*.

$$P = 2 \cdot \text{length} + 2 \cdot \text{width}$$

Use this formula to find the area of a rectangle.

$$A = \text{length} \cdot \text{width}$$

Area is measured in **square units**.

Find the perimeter and area of this rectangle.

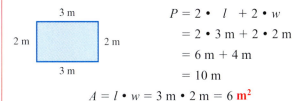

$$\begin{aligned} P &= 2 \cdot l + 2 \cdot w \\ &= 2 \cdot 3\text{ m} + 2 \cdot 2\text{ m} \\ &= 6\text{ m} + 4\text{ m} \\ &= 10\text{ m} \end{aligned}$$

$A = l \cdot w = 3\text{ m} \cdot 2\text{ m} = 6 \textbf{ m}^2$

Use these formulas to find the perimeter and area of a *square*.

$$P = 4 \cdot \text{side}$$
$$A = (\text{side})^2$$

Area is measured in **square units**.

Find the perimeter and area of this square.

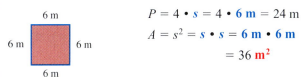

$P = 4 \cdot s = 4 \cdot \textbf{6 m} = 24\text{ m}$
$A = s^2 = s \cdot s = \textbf{6 m} \cdot \textbf{6 m}$
$ = 36 \textbf{ m}^2$

8.4 Parallelograms

Use these formulas to find the perimeter and area of a *parallelogram*.

$$P = \text{sum of the lengths of the sides}$$
$$A = \text{base} \cdot \text{height}$$

Area is measured in **square units**.

Find the perimeter and area of this parallelogram.

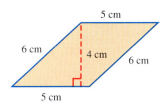

$P = 5\text{ cm} + 6\text{ cm} + 5\text{ cm} + 6\text{ cm} = 22\text{ cm}$
$A = 5\text{ cm} \cdot 4\text{ cm} = 20 \textbf{ cm}^2$

8.4 Trapezoids

Use these formulas to find the perimeter and area of a *trapezoid*.

$$P = \text{sum of the lengths of the sides}$$
$$A = \frac{1}{2} \cdot \text{height} \cdot (b + B)$$

where b is the short base and B is the long base.

Area is measured in **square units**.

Find the perimeter and area of this trapezoid.

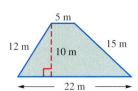

$P = 5\text{ m} + 15\text{ m} + 22\text{ m} + 12\text{ m} = 54\text{ m}$

$$A = \frac{1}{\cancel{2}} \cdot \cancel{10}^{5}\text{ m} \cdot (5\text{ m} + 22\text{ m})$$

$= 5\text{ m} \cdot (27\text{ m}) = 135 \textbf{ m}^2$

8.5 Triangles

Use these formulas to find the perimeter and area of a *triangle*.

$$P = \text{sum of the lengths of the sides}$$
$$A = \frac{1}{2} \cdot \text{base} \cdot \text{height}$$

or $\quad A = 0.5 \cdot \text{base} \cdot \text{height}$

Area is measured in **square units**.

Find the perimeter and area of this triangle.

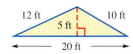

$P = 12\text{ ft} + 10\text{ ft} + 20\text{ ft} = 42\text{ ft}$

$$A = \frac{1}{2} \cdot b \cdot h$$

$$A = \frac{1}{\cancel{2}} \cdot \cancel{20}^{10}\text{ ft} \cdot 5\text{ ft} = 50 \textbf{ ft}^2$$

or $\quad A = 0.5 \cdot 20\text{ ft} \cdot 5\text{ ft} = 50 \textbf{ ft}^2$

Concepts	Examples
8.6 Circles Use this formula to find the *diameter* of a circle, given the radius. $$\text{diameter} = 2 \cdot \text{radius}$$	Find the diameter of a circle if the radius is 3 m. $$d = 2 \cdot r = 2 \cdot 3 \text{ m} = 6 \text{ m}$$
Use this formula to find the *radius* of a circle, given the diameter. $$\text{radius} = \frac{\text{diameter}}{2}$$	Find the radius of a circle if the diameter is 5 cm. $r = \dfrac{d}{2} = \dfrac{5 \text{ cm}}{2} = 2.5 \text{ cm}$
Use these formulas to find the **circumference** of a circle. $$C = 2 \cdot \pi \cdot \text{radius}$$ or $\quad C = \pi \cdot \text{diameter}$ Use 3.14 as the approximate value for π.	Find the circumference of a circle with a radius of 3 cm. $C = 2 \cdot \pi \cdot r$ $C \approx 2 \cdot 3.14 \cdot 3 \text{ cm} \approx 18.8 \text{ cm}$ **Rounded**
Use this formula to find the *area* of a circle. $$A = \pi \cdot (\text{radius})^2$$ Area is measured in **square units**.	Find the area of this circle. $A = \pi \cdot r^2$ $A \approx 3.14 \cdot 3 \text{ cm} \cdot 3 \text{ cm}$ $A \approx 28.3 \text{ cm}^2$ **Rounded**
8.7 Volume of a Rectangular Solid Use this formula to find the volume of *rectangular solids* (box-like shapes). $$V = \text{length} \cdot \text{width} \cdot \text{height}$$ Volume is measured in **cubic units**.	Find the volume of this box. 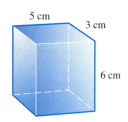 $V = l \cdot w \cdot h$ $V = 5 \text{ cm} \cdot 3 \text{ cm} \cdot 6 \text{ cm}$ $V = 90 \text{ cm}^3$
8.7 Volume of a Sphere and Hemisphere Use this formula to find the volume of a *sphere* (a ball-shaped solid). $$V = \frac{4}{3} \cdot \pi \cdot r^3$$ or $\quad V = \dfrac{4 \cdot \pi \cdot r^3}{3}$ where r is the radius of the sphere. Volume is measured in **cubic units**.	Find the volume of a sphere with a radius of 5 m. $V = \dfrac{4 \cdot \pi \cdot (\text{radius})^3}{3}$ $V \approx \dfrac{4 \cdot 3.14 \cdot 5 \text{ m} \cdot 5 \text{ m} \cdot 5 \text{ m}}{3}$ $V \approx 523.3 \text{ m}^3$ **Rounded**

(*continued*)

Concepts	Examples

8.7 Volume of a Sphere and Hemisphere (continued)

Use this formula to find the volume of a *hemisphere* (half of a sphere).

$$V = \frac{2}{3} \cdot \pi \cdot r^3$$

or $V = \dfrac{2 \cdot \pi \cdot r^3}{3}$

where r is the radius of the hemisphere.

Volume is measured in **cubic units**.

Find the volume of a hemisphere with a radius of 20 cm.

$$V = \frac{2 \cdot \pi \cdot (\text{radius})^3}{3}$$

$$V \approx \frac{2 \cdot 3.14 \cdot 20 \text{ cm} \cdot 20 \text{ cm} \cdot 20 \text{ cm}}{3}$$

$V \approx 16{,}746.7 \text{ cm}^3$ Rounded

8.7 Volume of a Cylinder

Use this formula to find the volume of a *cylinder*.

$$V = \pi \cdot r^2 \cdot h$$

where r is the radius of the circular base and h is the height of the cylinder.

Volume is measured in **cubic units**.

Find the volume of a cylinder that is 10 m high with a diameter of 8 m.

First, find the radius. $r = \dfrac{8 \text{ m}}{2} = 4 \text{ m}$

$V = \pi \cdot r^2 \cdot h$

$V \approx 3.14 \cdot 4 \text{ m} \cdot 4 \text{ m} \cdot 10 \text{ m}$

$V \approx 502.4 \text{ m}^3$

8.7 Volume of a Cone

Use this formula to find the volume of a *cone*.

$$V = \frac{1}{3} \cdot B \cdot h$$

or $V = \dfrac{B \cdot h}{3}$

where B is the area of the circular base and h is the height of the cone.

Volume is measured in **cubic units**.

Find the volume of a cone, with a height of 9 in. and a base with a radius of 4 in.

area of circular **B**ase $\approx 3.14 \cdot 4 \text{ in.} \cdot 4 \text{ in.}$

$B \approx 50.24 \text{ in.}^2$

$V = \dfrac{B \cdot h}{3}$

$V \approx \dfrac{50.24 \text{ in.}^2 \cdot 9 \text{ in.}}{3}$

$V \approx 150.7 \text{ in.}^3$ Rounded

8.7 Volume of a Pyramid

Use this formula to find the volume of a *pyramid*.

$$V = \frac{1}{3} \cdot B \cdot h$$

or $V = \dfrac{B \cdot h}{3}$

where B is the area of the square or rectangular base and h is the height of the pyramid.

Volume is measured in **cubic units**.

Find the volume of a pyramid with a square base 2 cm by 2 cm and a height of 6 cm.

area of square **B**ase $= 2 \text{ cm} \cdot 2 \text{ cm}$

$B = 4 \text{ cm}^2$

$V = \dfrac{B \cdot h}{3}$

$V = \dfrac{4 \text{ cm}^2 \cdot 6 \text{ cm}}{3}$

$V = 8 \text{ cm}^3$

8.8 Finding the Square Root of a Number

Use the square root key on a calculator, √ or √x .
Round to the nearest thousandth if necessary.

$\sqrt{64} = 8$ A perfect square

$\sqrt{43} \approx 6.557$ 6.557438524 is rounded to nearest thousandth

Concepts

8.8 Finding the Unknown Length in a Right Triangle

To find the *hypotenuse*, use this formula.

$$\text{hypotenuse} = \sqrt{(\text{leg})^2 + (\text{leg})^2}$$

The hypotenuse is the side opposite the right angle; it is the longest side in a right triangle.

To find a *leg*, use this formula.

$$\text{leg} = \sqrt{(\text{hypotenuse})^2 - (\text{leg})^2}$$

The legs are the sides that form the right angle.

8.9 Finding the Unknown Lengths in Similar Triangles

Use the fact that in similar triangles, the ratios of the lengths of corresponding sides are equal. Write a proportion. Then find the cross products and show that they are equivalent. Finish solving for the unknown length.

Examples

Find the unknown length in this right triangle. Round to the nearest tenth.

$$\text{hypotenuse} = \sqrt{(6)^2 + (5)^2}$$
$$= \sqrt{36 + 25}$$
$$= \sqrt{61} \approx 7.8$$

The hypotenuse is about 7.8 m long.

Find the unknown length in this right triangle. Round to the nearest tenth.

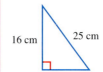

$$\text{leg} = \sqrt{(25)^2 - (16)^2}$$
$$= \sqrt{625 - 256}$$
$$= \sqrt{369} \approx 19.2$$

The leg is about 19.2 cm long.

Find x and y if the triangles are similar.

$$\frac{x}{8} = \frac{5}{10} \qquad\qquad \frac{y}{12} = \frac{5}{10}$$

$$x \cdot 10 = 8 \cdot 5 \qquad\qquad y \cdot 10 = 12 \cdot 5$$

$$\frac{x \cdot \cancel{10}}{\cancel{10}} = \frac{\cancel{40}}{\cancel{10}} \qquad\qquad \frac{y \cdot \cancel{10}}{\cancel{10}} = \frac{\cancel{60}}{\cancel{10}}$$

$$x = 4 \text{ m} \qquad\qquad y = 6 \text{ m}$$

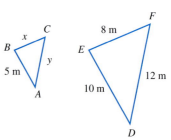

ANSWERS TO TEST YOUR WORD POWER

1. (d) *Example:* If ∠1 measures 35° and ∠2 measures 55°, the angles are complementary because 35° + 55° = 90°.
2. (b) *Example:* If a square measures 5 ft on each side, then the perimeter is 5 ft + 5 ft + 5 ft + 5 ft = 20 ft.
3. (c) *Examples:* Angles that measure 91°, 120°, and 175° are all obtuse angles.   **4. (c)** *Example:* In triangle *ABC* at the right, side *AC* is the hypotenuse; sides *AB* and *BC* are the legs.   **5. (b)** *Example:* The ratio of a circumference of 12.57 cm to a diameter of 4 cm is $\frac{12.57}{4} \approx 3.14$ (rounded).   **6. (a)** *Example:* $\overleftrightarrow{EF}$ is perpendicular to $\overleftrightarrow{GH}$, at the right.   **7. (d)** *Example:* Triangle *ABC* is similar to triangle *DEF*, so the ratios of corresponding sides are equal. $\frac{AB}{DE} = \frac{3 \text{ m}}{6 \text{ m}} = \frac{1}{2} \quad \frac{BC}{EF} = \frac{2 \text{ m}}{4 \text{ m}} = \frac{1}{2} \quad \frac{AC}{DF} = \frac{3.5 \text{ m}}{7 \text{ m}} = \frac{1}{2}$

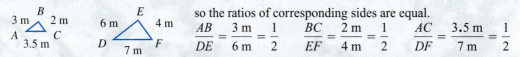

8. (b) *Example:* In a rectangle with a length of 8 in. and a width of 5 in., Area = 8 in. • 5 in. = 40 in.²

Chapter 8 REVIEW EXERCISES

[8.1] *Name each line, line segment, or ray.*

1.
2.
3.

Label each pair of lines as parallel, perpendicular, or intersecting.

4.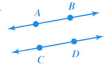
5.
6.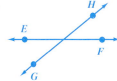

Label each angle as acute, right, obtuse, or straight. For right and straight angles, indicate the number of degrees in the angle.

7.
8.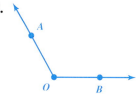

9.
10.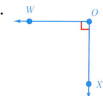

[8.2] *In each figure you are given the measures of two of the angles. Find the measure of each of the other angles.*

11.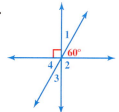
12.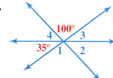

Name the pairs of supplementary angles in each figure.

13.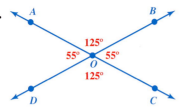

14.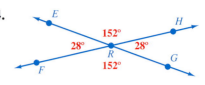

Find the complement or supplement of each angle.

15. Find the complement of:
    (a) 80°
    (b) 45°
    (c) 7°

16. Find the supplement of:
    (a) 155°
    (b) 90°
    (c) 33°

[8.3] *Find the perimeter of each rectangle or square.*

17.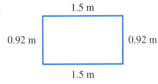

18.

19. A square-shaped pillow measures 38 cm along each side. How much lace is needed to trim all the edges?

20. A rectangular garden plot is $8\frac{1}{2}$ ft wide and 12 ft long. How much fencing is needed to surround the garden?

Find the area of each rectangle or square. Round your answers to the nearest tenth when necessary.

21.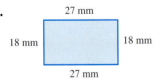

22.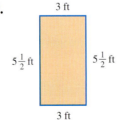

23.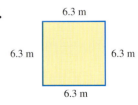

[8.4] *Find the perimeter and area of each parallelogram or trapezoid. Round your answers to the nearest tenth when necessary.*

24.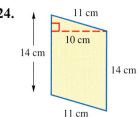

25.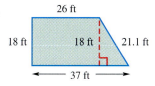

26.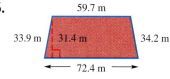

[8.5] *Find the perimeter and area of each triangle.*

27.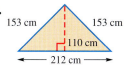

28.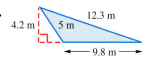

29.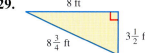

Find the number of degrees in each unlabeled angle.

30.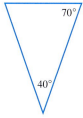

31.

[8.6] *Find the unknown length.*

32. The radius of a circular irrigation field is 68.9 m. What is the diameter of the field?

33. The diameter of a juice can is 3 in. What is the radius of the can?

Find the circumference and area of each circle. Use 3.14 as the approximate value for π. Round your answers to the nearest tenth.

34.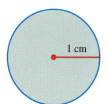

35.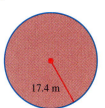

36.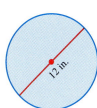

606 Chapter 8 Geometry

[8.3–8.6] *Find each shaded area. Use 3.14 as the approximate value for π. Round your answers to the nearest tenth when necessary.*

37.

38.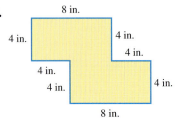

39.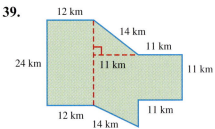

40.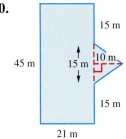

41.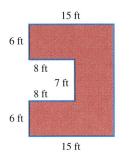

42.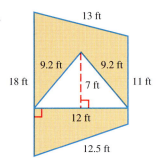

43.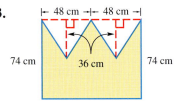

44.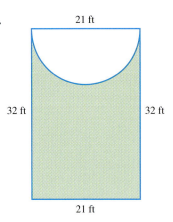

45.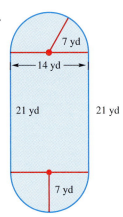

[8.7] *Find each volume. Use 3.14 as the approximate value for π. Round your answers to the nearest tenth when necessary.*

46.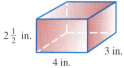

47.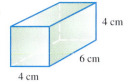

48.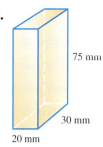

49.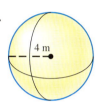

50.

51.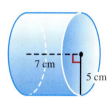

52.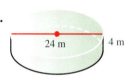

53.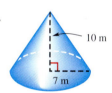

54.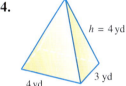

[8.8] *Find each square root. Round your answers to the nearest thousandth when necessary.*

55. $\sqrt{49}$

56. $\sqrt{8}$

57. $\sqrt{3000}$

58. $\sqrt{144}$

59. $\sqrt{58}$

60. $\sqrt{625}$

61. $\sqrt{105}$

62. $\sqrt{80}$

Find the unknown length in each right triangle. Use a calculator to find square roots. Round your answers to the nearest tenth when necessary.

63.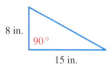

64.

65.

66.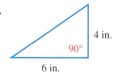

67.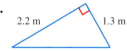

68.

608 Chapter 8 Geometry

[8.9] *Find the unknown lengths in each pair of similar triangles. Then find the perimeter of the larger triangle in each pair.*

69.

70.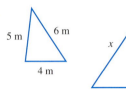

71.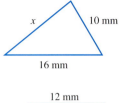

MIXED REVIEW EXERCISES

Find the perimeter (or circumference) and area of each figure. Use 3.14 *as the approximate value for* π. *Round your answers to the nearest tenth when necessary.*

72.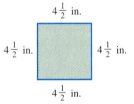

73.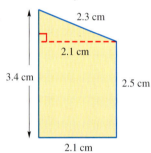

74.

75.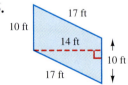

76.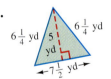

77.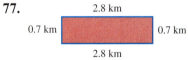

78.

79.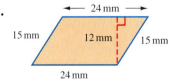

80.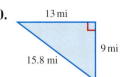

Label each figure. Choose from these labels: line segment, ray, parallel lines, perpendicular lines, intersecting lines, acute angle, right angle, straight angle, obtuse angle. Indicate the number of degrees in the right angle and the straight angle.

81.

82.

83.

84.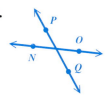

85.

86.

87.

88.

89.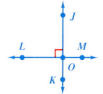

90. What is the complement of an angle measuring 9°?

91. What is the supplement of an angle measuring 42°?

Find the perimeter and area of each figure.

92.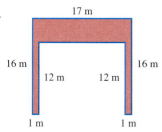

93.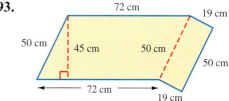

Find the volume of each figure. Use **3.14** *as the approximate value for* π. *Round your answers to the nearest tenth when necessary.*

94.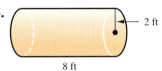

95.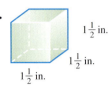

96.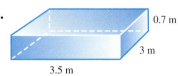

97.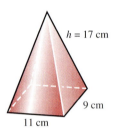

98.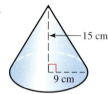

99.

Find the unknown angle or side measurement. Round your answers to the nearest tenth when necessary.

100.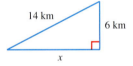

101.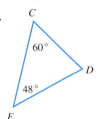

102. similar triangles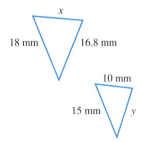

103. Explain how you could use the information about prefixes from **Section 8.6** to solve a problem that asks, "How many decades are in two centuries?"

Chapter 8 Test

Choose the figure that matches each label. For right and straight angles, indicate the number of degrees in the angle.

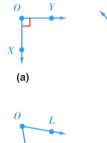

(a) (b) (c) (d)

(e) (f) (g)

1. Acute angle is figure _____.

2. Right angle is figure _____ and its measure is _____.

3. Ray is figure _____.

4. Straight angle is figure _____ and its measure is _____.

5. Write a definition of parallel lines and a definition of perpendicular lines. Make a sketch to illustrate each definition.

6. Find the complement of an 81° angle.

7. Find the supplement of a 20° angle.

8. Find the measure of each unlabeled angle in the figure at the right.

Find the perimeter and area of each figure.

9.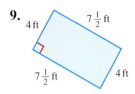
 4 ft, $7\frac{1}{2}$ ft, $7\frac{1}{2}$ ft, 4 ft

10.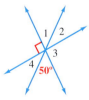
 18 mm, 18 mm, 18 mm, 18 mm

11.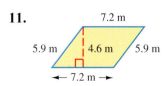
 7.2 m, 5.9 m, 4.6 m, 5.9 m, 7.2 m

12.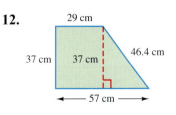
 29 cm, 37 cm, 37 cm, 46.4 cm, 57 cm

1. _____
2. _____
3. _____
4. _____
5. _____
6. _____
7. _____
8. _____
9. _____
10. _____
11. _____
12. _____

Find the perimeter and area of each triangle.

13. 11.8 m, 8 m, 8.65 m, 12 m

14. 9 yd, 13 yd, $15\frac{4}{5}$ yd

15. A triangle has angles that measure 90° and 35°. What does the third angle measure?

In Problems 16–22, use 3.14 as the approximate value for π. Round your answers to the nearest tenth when necessary.

16. Find the radius.
    25 in.

17. Find the circumference.
    0.9 km

Find the area of each figure.

18. 16.2 cm

19. 5 m

Find the volume of each figure.

20. 12 m, 30 m, 18 m

21. 2 ft

22. 5 ft, 18 ft

Find the unknown lengths. Round your answers to the nearest tenth when necessary.

23. ?, 6 cm, 7 cm

24. Similar triangles
    18 cm, 15 cm, 9 cm, 10 cm, z, y

25. Explain the difference between cm, cm^2, and cm^3. In what types of geometry problems might you use each of these units?

Cumulative Review Exercises — Chapters 1–8

First use front end rounding to round each number and estimate the answer. Then find the exact answer.

1. Estimate: Exact:
   ____   319
                  58,028
   + ____   + 6 227

2. Estimate: Exact:
                   20.07
   − ____   − 9.828

3. Estimate: Exact:
                   3.664
   × ____   ×   7.3

4. Estimate: Exact:
                   28,419
   × ____   ×   73

5. Estimate: Exact:
   ____)____   2.8)562.24

6. Estimate: Exact:
   ____)____   52)4888

7. Estimate: Exact:
                   $4\frac{1}{2}$
   + ____   + $4\frac{9}{10}$

8. Estimate: Exact:
                   $3\frac{1}{6}$
   − ____   − $1\frac{7}{8}$

9. Exact:
   $3\frac{1}{9} \cdot 1\frac{5}{7} = $ _____

   Estimate:
   ___ · ___ = ___

Add, subtract, multiply, or divide as indicated. Write answers to fraction problems in lowest terms and as whole or mixed numbers when possible.

10. $3\frac{3}{5} \div 8$

11. $1 - 0.0868$

12. Write your answer using R for the remainder.

    $81\overline{)5749}$

13. $10 \div \frac{5}{16}$

14. $(0.006)(0.013)$

15. $40{,}020 - 915$

16. $0.7 \div 0.036$ Round answer to nearest hundredth.

17. $6\frac{1}{6} - 1\frac{3}{4}$

18. $752.6 + 83 + 0.485$

Use the order of operations to simplify each expression.

19. $16 - (10 - 2) \div 2 \cdot 3 + 5$

20. $2^4 \div \sqrt{64} + 6^2$

21. Write 0.0208 in words.

22. Write six hundred sixty and five hundredths in numbers.

Chapter 8 Geometry

23. Arrange in order from smallest to largest.
2.55   2.505   2.055   2.5005

24. Explain how you could use the information on prefixes and root words in Section 8.6 to remember the way to change a percent to a decimal.

Complete this chart.

Fraction/Mixed Number	Decimal	Percent
25. _____	0.02	26. _____
$1\frac{3}{4}$	27. _____	28. _____
29. _____	30. _____	40%

Write each rate or ratio as a fraction in lowest terms. Change to the same units when necessary.

31. 4 ft to 6 in.

32. Last month there were 9 cloudy days and 21 sunny days. What was the ratio of sunny days to cloudy days?

Find the unknown number in each proportion. Round your answer to hundredths if necessary.

33. $\dfrac{5}{13} = \dfrac{x}{91}$

34. $\dfrac{207}{69} = \dfrac{300}{x}$

35. $\dfrac{4.5}{x} = \dfrac{6.7}{3}$

Solve each percent problem.

36. 72 patients is what percent of 45 patients?

37. $18 is 3% of what number of dollars?

Convert each measurement.

38. $2\frac{1}{4}$ hours to minutes

39. 40 oz to pounds

40. 8 cm to meters

41. 1.8 L to milliliters

Write the most reasonable metric unit in each blank. Choose from km, m, cm, mm, L, mL, kg, g, and mg.

42. Her wristwatch strap is 15 _____ wide.

43. Jon added 2 _____ of oil to his car.

44. The child weighs 15 _____.

45. The bookcase is 90 _____ high.

46. List the metric temperatures at which water freezes and water boils.

Cumulative Review Exercises: Chapters 1–8 **615**

Find the perimeter (or circumference) and area of each figure. Use 3.14 as the approximate value of π.

47. 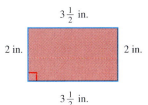   **48.** 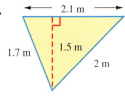   **49.**

50. 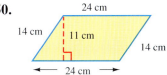   **51.** 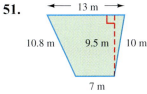   **52.**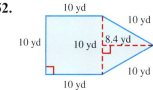

Find the unknown length in each figure. Round your answers to the nearest tenth. In Exercise 54, also find the perimeter of the smaller triangle.

53.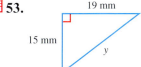

54. Similar triangles

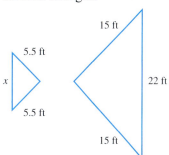

Solve each application problem.

55. Mei Ling must earn 90 credits to receive an associate of arts degree. She has 53 credits. What percent of the necessary credits does she have? Round to the nearest whole percent.

56. Which bag of chips is the best buy: Brand T is $15\frac{1}{2}$ oz for \$2.99, Brand F is 14 oz for \$2.49, and Brand H is 18 oz for \$3.89. You have a coupon for 40¢ off Brand H and another for 30¢ off Brand T.

57. A Folger's coffee can has a diameter of 13 cm and a height of 17 cm. Find the volume of the can. Use 3.14 for π and round your answer to the nearest tenth. (*Source:* Folger's.)

58. A photograph measures 8 in. by 10 in. Earl put it in a frame that is 2 in. wide. Find the perimeter of the frame. To help solve this problem, first label the inside and outside measurements on the sketch of the frame.

59. Steven bought $4\frac{1}{2}$ yd of canvas material to repair the tents used by the scout troop. He used $1\frac{2}{3}$ yd on one tent and $1\frac{3}{4}$ yd on another. How much material is left?

60. The cooks at a homeless shelter used 30 lb of meat to make stew for 140 people. At that rate, how much meat is needed for stew to feed 200 people? Round to the nearest tenth.

61. Graciela needs 85 cm of yarn to make a tassel for one corner of a pillow. How many meters of yarn does she need to put a tassel on each corner of a square-shaped pillow?

62. Swimsuits are on sale in August at 65% off the regular price. How much will Lanece pay for a suit that has a regular price of $64?

The table shows the changes in U.S. postal rates that went into effect in January 2001. Use the information in the table to answer Exercises 63–66. Write all money answers using a dollar sign an decimal point.

CHANGES IN U.S. POSTAL RATES, JANUARY 2001

Item	From	To
First class, 1st ounce	33¢	34¢
First class, 2nd–11th ounce	22¢/oz	21¢/oz
Postcard	20¢	20¢
Priority mail up to 2 lbs	$3.20	$3.95
Newspapers up to 10 oz	26.6¢	28.7¢

Source: Associated Press.

63. How much would it cost under the old rates and under the new rates to mail a first class envelope weighing **(a)** 2 oz, **(b)** 6 oz, **(c)** 11 oz?

64. Find the percent increase or decrease, to the nearest tenth of a percent, in the rates for **(a)** first class, 1st ounce; **(b)** first class, 2nd ounce; **(c)** postcards; **(d)** priority mail.

65. Can a 36 oz package be sent by priority mail for $3.95 under the new rates? Explain your answer.

66. The *Wall Street Journal* has the largest circulation of any newspaper in the United States, with a daily average of 1,762,750. (*Source:* Audit Bureau of Circulation.) If 12% of the papers are sent by mail, and each paper weighs less than 10 oz, what is the daily increase in postage costs from the old to new rates?

Answers to Selected Exercises **A-27**

28. $\frac{1}{20}$ **29.** 5% **30.** $3\frac{1}{2}$ **31.** 3.5 **32.** $\frac{1}{3}$ **33.** $\frac{8}{1}$ **34.** $\frac{8}{3}$
35. 12 **36.** 1.67 (rounded) **37.** 5% **38.** 30 hours **39.** 30 in.
40. $1\frac{3}{4}$ or 1.75 min **41.** 280 cm **42.** 0.065 g **43.** 122.76 mi
44. 10 °C **45.** g **46.** cm **47.** m **48.** kg **49.** mg **50.** mL
51. 38 °C **52.** 12 °C **53.** $18.99 (rounded); $170.95
54. $0.46 (rounded) **55.** $12\frac{3}{4}$ or 12.75 lb
56. 93.6 km; 58.0 mi (rounded) **57.** $2756.25; $11,506.25
58. 88.6% (rounded) **59.** $9.74 (rounded)
60. $\frac{1}{6}$ cup more than the amount needed **61.** 120 rows
62. 2100 students **63.** (a) 165.5 hr (b) 8.1% (rounded)
64. (a) 8.5 hr (b) 4.7% (rounded)
65. (a) $1.3 million (rounded) (b) $2.5 million (rounded)
(c) $4.2 million (rounded)

Chapter 8

Section 8.1 (page 527)
1. line named $\overleftrightarrow{CD}$ or $\overleftrightarrow{DC}$ **3.** line segment named $\overline{GF}$ or $\overline{FG}$
5. ray named $\overrightarrow{PQ}$ **7.** perpendicular **9.** parallel **11.** intersecting **13.** $\angle AOS$ or $\angle SOA$ **15.** $\angle CRT$ or $\angle TRC$ **17.** $\angle AQC$ or $\angle CQA$ **19.** right (90°) **21.** acute **23.** straight (180°)
25. (a) The car turned around in a complete circle. (b) The governor took the opposite view, for example, having once opposed taxes but now supporting them. **27.** True, because $\overleftrightarrow{UQ}$ is perpendicular to $\overleftrightarrow{ST}$. **28.** True, because they form a 90° angle, as indicated by the small red square. **29.** False; the angles have the same measure (both are 180°). **30.** False; $\overleftrightarrow{ST}$ and $\overleftrightarrow{PR}$ are parallel. **31.** False; $\overleftrightarrow{QU}$ and $\overleftrightarrow{TS}$ are perpendicular. **32.** True; both angles are formed by perpendicular lines, so they both measure 90°.

Section 8.2 (page 533)
1. $\angle EOD$ and $\angle COD$; $\angle AOB$ and $\angle BOC$ **3.** $\angle HNE$ and $\angle ENF$; $\angle HNG$ and $\angle GNF$; $\angle HNE$ and $\angle HNG$; $\angle ENF$ and $\angle GNF$
5. 50° **7.** 4° **9.** 50° **11.** 90° **13.** $\angle SON \cong \angle TOM$; $\angle TOS \cong \angle MON$ **15.** $\angle GOH$ measures 63°; $\angle EOF$ measures 37°; $\angle AOC$ and $\angle GOF$ both measure 80°. **17.** Two angles are complementary if their sum is 90°. Two angles are supplementary if their sum is 180°. Drawings will vary; examples are:

Complimentary

Supplementary

19. $\angle ABF \cong \angle ECD$; Both are 138°. $\angle ABC \cong \angle BCD$; Both are 42°.
21. No; obtuse angles are >90°, so their sum would be >180°.

Section 8.3 (page 541)
1. $P = 28$ yd; $A = 48$ yd^2 **3.** $P = 3.6$ km; $A = 0.81$ km^2
5. $P = 40$ ft; $A = 100$ ft^2 **7.** $P = 29$ m; $A = 7$ m^2
9. $P = 196.2$ ft; $A = 1674.2$ ft^2 **11.** $P = 12$ mi; $A = 9$ mi^2
13. $P = 38$ m; $A = 39$ m^2 **15.** $P = 98$ m; $A = 492$ m^2
17. unlabeled side = 12 in.; $P = 78$ in.; $A = 234$ in.^{2} **19.** $460
21. $94.81 **23.** Panoramic: $P = 36$ in., $A = 56$ in.^{2}; 4 in. × 6 in.: $P = 20$ in., $A = 24$ in.^{2}; 4 in. × 7 in.: $P = 22$ in., $A = 28$ in.^{2}
25. 53 yd

27. $A = 6528$ ft^2

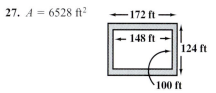

29.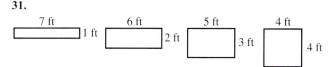

30. (a) 5 ft by 1 ft has area of 5 ft^2; 4 ft by 2 ft has area of 8 ft^2; 3 ft by 3 ft has area of 9 ft^2 (b) The square plot 3 ft by 3 ft has greatest area.
31.

| 7 ft × 1 ft | 6 ft × 2 ft | 5 ft × 3 ft | 4 ft × 4 ft |

32. (a) 7 ft^2; 12 ft^2; 15 ft^2; 16 ft^2 (b) Square plots have the greatest area.
33. (a) $P = 10$ ft; $A = 6$ ft^2 (3 ft × 2 ft)
(b) $P = 20$ ft; $A = 24$ ft^2 (6 ft × 4 ft)
(c) Perimeter is twice the original; area is four times the original.
34. (a) $P = 30$ ft; $A = 54$ ft^2 (9 ft × 6 ft)

(b) Perimeter is three times the original; area is nine times the original. (c) Perimeter will be four times the original; area will be 16 times the original.

Section 8.4 (page 549)
1. 208 m **3.** 207.2 m **5.** 5.78 km **7.** 775 mm^2 **9.** 19.25 ft^2
11. 3099.6 cm^2 **13.** $1410.75 **15.** 437.5 in.^{2} **17.** Height is not part of perimeter; square units are used for area, not perimeter. $P = 2.5$ cm + 2.5 cm + 2.5 cm + 2.5 cm = 10 cm **19.** 3.02 m^2
21. 25,344 ft^2

Section 8.5 (page 555)
1. $P = 202$ m; $A = 1914$ m^2 **3.** $P = 58.9$ cm; $A = 139.15$ cm^2
5. $P = 26\frac{1}{4}$ yd; $A = 30\frac{3}{4}$ yd^2 **7.** $P = 85.2$ cm; $A = 302.46$ cm^2
9. $A = 198$ m^2 **11.** $A = 1664$ m^2 **13.** 32° **15.** 48°
17. No. Right angles are 90°, so two right angles are 180°, and the sum of all *three* angles in a triangle equals 180°. **19.** $7\frac{7}{8}$ ft^2 or 7.875 ft^2 **21.** 126.8 m of curb; 672 m^2 of sod **23.** (a) 32 m^2
(b) 13.5 m^2

Section 8.6 (page 565)
1. $d = 18$ mm **3.** $r = 0.35$ km **5.** $C \approx 69.1$ ft; $A \approx 379.9$ ft^2
7. $C \approx 8.2$ m; $A \approx 5.3$ m^2 **9.** $C \approx 47.1$ cm; $A \approx 176.6$ cm^2
11. $C \approx 23.6$ ft; $A \approx 44.2$ ft^2 **13.** $C \approx 27.2$ km; $A \approx 58.7$ km^2

15. $A \approx 57$ cm² **17.** $A \approx 197.8$ cm² **19.** π is the ratio of circumference of a circle to its diameter. If you divide the circumference of any circle by its diameter, the answer is always a little more than 3. The approximate value is 3.14, which we call π (pi). Your test question could involve finding the circumference or the area of a circle. **21.** $C \approx 219.8$ cm **23.** $C \approx 785.0$ ft **25.** $A \approx 70{,}650$ mi² **27.** watch: $C \approx 3.1$ in., $A \approx 0.8$ in.²; wall clock: $C \approx 18.8$ in., $A \approx 28.3$ in.² **29.** $d \approx 45.9$ cm **31.** $1170.33 (rounded) **33.** The prefix *rad* tells you that radius is a ray from the center of the circle. The prefix *dia* means the diameter goes through the circle, and the prefix *circum* means the circumference is the distance around. **35.** $A \approx 44.2$ in.² **36.** $A \approx 132.7$ in.² **37.** $A \approx 201.0$ in.² **38.** small: $0.063 (rounded); medium: $0.049 (rounded); large: $0.046 (rounded) Best Buy **39.** small: $0.084 (rounded); medium: $0.067 (rounded) Best Buy; large: $0.071 (rounded) **40.** small: $0.077 (rounded) Best Buy; medium: $0.083 (rounded); large: $0.078 (rounded)

Summary Exercises on Perimeter, Circumference, and Area (page 569)

1. (a) All sides have the same length. **(b)** Opposite sides have the same length.

(c) Opposite sides have the same length.

2. (a) **(b)**

(c) The diameter is twice the radius, or the radius is half the diameter. **3.** Add up the lengths of all the sides to find the perimeter. **4.** Perimeter is the total distance around the outside edges of a shape. Area is the number of square units needed to cover the space inside the shape. **5.** f, e, d, b, c, a **6. (a)** $C = 2 \cdot \pi \cdot r$ **(b)** $C = \pi \cdot d$ **(c)** Divide the diameter by 2 to find the radius. **7.** $P = 63$ ft; $A = 180$ ft² $P = 32.4$ cm; $A \approx 45.1$ cm²

8. (a) $A = 12$ cm² **(b)** $P = 6\frac{1}{2}$ ft **(c)** $C \approx 28.5$ m **(d)** $A = 307$ in.²

9. $P = 27$ in.; $A = 31\frac{1}{2}$ in.² or 31.5 in.² **10.** $P = 48$ yd; $A = 144$ yd² **11.** $P = 6.4$ m; $A \approx 2.2$ m² **12.** $P = 33.5$ mm; $A = 64$ mm² **13.** $d = 12$ cm; $C \approx 37.7$ cm; $A \approx 113.0$ cm² **14.** $r = 15$ mi; $C \approx 94.2$ mi; $A \approx 706.5$ mi² **15.** $r = 4.5$ ft; $C \approx 28.3$ ft; $A \approx 63.6$ ft² **16.** $A \approx 4.5$ m² **17.** $A \approx 217.5$ yd² **18.** $A \approx 86$ in.² **19.** $C \approx 14.6$ ft; Bonus: About 361.6 revolutions; the Mormons used 360, which is the answer you get using $2\frac{1}{3}$ ft instead of 2.33 ft as the radius. **20.** $4.47 (rounded)

Section 8.7 (page 577)

1. $V = 550$ cm³ **3.** $V \approx 44{,}579.6$ m³ **5.** $V \approx 3617.3$ in.³ **7.** $V \approx 471$ ft³ **9.** $V \approx 418.7$ m³ **11.** $V = 800$ cm³

13. $V = 18$ in.³ **15.** $V \approx 2481.5$ cm³ **17.** $V = 651{,}775$ m³ **19.** $V \approx 3925$ ft³ **21.** Student used diameter of 7 cm; should use radius of 3.5 cm in formula. Units for volume are cm³, not cm². Correct answer is $V \approx 192.3$ cm³. **23.** $V = 513$ cm³

Section 8.8 (page 583)

1. 4 **3.** 8 **5.** 3.317 **7.** 2.236 **9.** 8.544 **11.** 10.050 **13.** 13.784 **15.** 31.623 **17.** 30 is about halfway between 25 and 36, so $\sqrt{30}$ should be about halfway between 5 and 6, or about 5.5. Using a calculator, $\sqrt{30} \approx 5.477$. Similarly, $\sqrt{26}$ should be a little more than $\sqrt{25}$; by calculator $\sqrt{26} \approx 5.099$. And $\sqrt{35}$ should be a little less than $\sqrt{36}$; by calculator $\sqrt{35} \approx 5.916$. **19.** $\sqrt{1521} = 39$ ft **21.** $\sqrt{289} = 17$ in. **23.** $\sqrt{144} = 12$ mm **25.** $\sqrt{73} \approx 8.5$ in. **27.** $\sqrt{65} \approx 8.1$ yd **29.** $\sqrt{195} \approx 14.0$ cm **31.** $\sqrt{7.94} \approx 2.8$ m **33.** $\sqrt{65.01} \approx 8.1$ cm **35.** $\sqrt{292.32} \approx 17.1$ km **37.** $\sqrt{65} \approx 8.1$ ft **39.** $\sqrt{360{,}000} = 600$ m **41.** $\sqrt{135} \approx 11.6$ ft

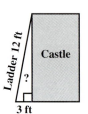

43. The student did not square the numbers correctly: 9^2 is 81 and 7^2 is 49. Also, the final answer is rounded to thousandths instead of tenths. Correct answer is $\sqrt{130} \approx 11.4$ in. **45.** $\sqrt{16200} \approx 127.3$ ft **46.** $\sqrt{7200} \approx 84.9$ ft

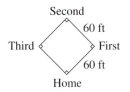

47. The distance from third to first is the same as the distance from home to second because the baseball diamond is a square. **48.** One possibility is:

$$\text{major league } \frac{90 \text{ ft}}{127.3 \text{ ft}} = \frac{60 \text{ ft}}{x} \text{ softball}$$

$x \approx 84.9$ ft

Section 8.9 (page 591)

1. similar **3.** not similar **5.** similar **7.** $\angle 1$ and $\angle 4$; $\angle 2$ and $\angle 5$; $\angle 3$ and $\angle 6$; $\overline{AB}$ and $\overline{PQ}$; $\overline{BC}$ and $\overline{QR}$; $\overline{AC}$ and $\overline{PR}$ **9.** $\angle 1$ and $\angle 6$; $\angle 2$ and $\angle 5$; $\angle 3$ and $\angle 4$; $\overline{MP}$ and $\overline{QS}$; $\overline{MN}$ and $\overline{QR}$; $\overline{NP}$ and $\overline{RS}$ **11.** $\frac{3}{2}; \frac{3}{2}; \frac{3}{2}$ **13.** $a = 5$ mm; $b = 3$ mm **15.** $a = 6$ cm; $b = 15$ cm **17.** $x = 24.8$ m; $P = 72.8$ m; $y = 15$ m; $P = 54.6$ m **19.** $P = 8$ cm + 8 cm + 8 cm = 24 cm; $A = 0.5 \cdot 8$ cm $\cdot 6.9$ cm $= 27.6$ cm² **21.** $h = 24$ ft **23.** $\frac{3}{4}$ ft or 0.75 ft **25.** One dictionary definition is "resembling, but not identical." Examples of similar objects are sets of different size pots or measuring cups; small and large size cans of beans; child's tennis shoe and adult's tennis shoe. **27.** $x = 50$ m **29.** $n = 110$ m

Chapter 8 Review Exercises (page 603)

1. line segment named $\overline{AB}$ or $\overline{BA}$ 2. line named $\overleftrightarrow{CD}$ or $\overleftrightarrow{DC}$
3. ray named $\overrightarrow{OP}$ 4. parallel 5. perpendicular 6. intersecting
7. acute 8. obtuse 9. straight; 180° 10. right; 90°
11. ∠1 and ∠3 measure 30°; ∠2 measures 90°; ∠4 measures 60°
12. ∠1 measures 100°; ∠2 and ∠4 measure 45°; ∠3 measures 35°
13. ∠AOB and ∠BOC; ∠BOC and ∠COD; ∠COD and ∠DOA; ∠DOA and ∠AOB 14. ∠ERH and ∠HRG; ∠HRG and ∠GRF; ∠FRG and ∠FRE; ∠FRE and ∠ERH 15. (a) 10° (b) 45° (c) 83°
16. (a) 25° (b) 90° (c) 147° 17. $P = 4.84$ m 18. $P = 128$ in.
19. 152 cm 20. 41 ft 21. $A = 486$ mm² 22. $A = 16.5$ ft² or $16\frac{1}{2}$ ft² 23. $A \approx 39.7$ m² 24. $P = 50$ cm; $A = 140$ cm²
25. $P = 102.1$ ft; $A = 567$ ft² 26. $P = 200.2$ m; $A \approx 2074.0$ m²
27. $P = 518$ cm; $A = 11,660$ cm² 28. $P = 27.1$ m; $A = 20.58$ m²
29. $P = 20\frac{1}{4}$ ft or 20.25 ft; $A = 14$ ft² 30. 70° 31. 24°
32. $d = 137.8$ m 33. $r = 1\frac{1}{2}$ in. or 1.5 in. 34. $C \approx 6.3$ cm; $A \approx 3.1$ cm² 35. $C \approx 109.3$ m; $A \approx 950.7$ m² 36. $C \approx 37.7$ in.; $A \approx 113.0$ in.² 37. $A \approx 20.3$ m² 38. $A = 64$ in.²
39. $A = 673$ km² 40. $A = 1020$ m² 41. $A = 229$ ft²
42. $A = 132$ ft² 43. $A = 5376$ cm² 44. $A \approx 498.9$ ft²
45. $A \approx 447.9$ yd² 46. $V = 30$ in.³ 47. $V = 96$ cm³
48. $V = 45,000$ mm³ 49. $V \approx 267.9$ m³ 50. $V \approx 452.2$ ft³
51. $V \approx 549.5$ cm³ 52. $V \approx 1808.6$ m³ 53. $V \approx 512.9$ m³
54. $V = 16$ yd³ 55. 7 56. 2.828 (rounded)
57. 54.772 (rounded) 58. 12 59. 7.616 (rounded)
60. 25 61. 10.247 (rounded) 62. 8.944 (rounded)
63. $\sqrt{289} = 17$ in. 64. $\sqrt{49} = 7$ cm 65. $\sqrt{104} \approx 10.2$ cm
66. $\sqrt{52} \approx 7.2$ in. 67. $\sqrt{6.53} \approx 2.6$ m
68. $\sqrt{71.75} \approx 8.5$ km 69. $y = 30$ ft; $x = 34$ ft; $P = 104$ ft
70. $y = 7.5$ m; $x = 9$ m; $P = 22.5$ m 71. $x = 12$ mm; $y = 7.5$ mm; $P = 38$ mm 72. $P = 18$ in; $A \approx 20.3$ in.² or $A = 20\frac{1}{4}$ in.² 73. $P = 10.3$ cm; $A \approx 6.2$ cm² 74. $C \approx 40.8$ m; $A \approx 132.7$ m² 75. $P = 54$ ft; $A = 140$ ft² 76. $P = 20$ yd; $A = 18\frac{3}{4}$ yd² or 18.8 yd² (rounded) 77. $P = 7$ km; $A \approx 2.0$ km²
78. $C \approx 53.4$ m; $A \approx 226.9$ m² 79. $P = 78$ mm; $A \approx 288$ mm²
80. $P = 37.8$ mi; $A = 58.5$ mi² 81. parallel lines 82. line segment 83. acute angle 84. intersecting lines 85. right angle; 90° 86. ray 87. straight angle; 180° 88. obtuse angle
89. perpendicular lines 90. 81° 91. 138° 92. $P = 90$ m; $A = 92$ m² 93. $P = 282$ cm; $A = 4190$ cm² 94. $V \approx 100.5$ ft³
95. $V \approx 3.4$ in.³ or $V = 3\frac{3}{8}$ in.³ 96. $V \approx 7.4$ m³
97. $V = 561$ cm³ 98. $V \approx 1271.7$ cm³ 99. $V \approx 1436.0$ m³
100. $x \approx 12.6$ km 101. $\angle D = 72°$ 102. $x = 12$ mm; $y = 14$ mm 103. The prefix *dec* in *dec*ade means 10 and the prefix *cent* in *cent*ury means 100, so divide 200 (two centuries) by 10. The answer is 20 decades.

Chapter 8 Test (page 611)

1. (e) 2. (a); 90° 3. (d) 4. (g); 180° 5. Parallel lines are lines in the same plane that never intersect. Perpendicular lines intersect to form a right angle.

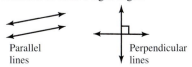

6. 9° 7. 160° 8. ∠1 measures 50°; ∠3 measures 90°; ∠2 and ∠4 measure 40° 9. $P = 23$ ft; $A = 30$ ft²
10. $P = 72$ mm; $A = 324$ mm² 11. $P = 26.2$ m; $A = 33.12$ m²
12. $P = 169.4$ cm; $A = 1591$ cm² 13. $P = 32.45$ m; $A = 48$ m²
14. $P = 37.8$ yd or $37\frac{4}{5}$ yd; $A = 58.5$ yd² 15. 55°
16. $r = 12.5$ in. or $12\frac{1}{2}$ in. 17. $C \approx 5.7$ km 18. $A \approx 206.0$ cm²
19. $A \approx 39.3$ m² 20. $V = 6480$ m³ 21. $V \approx 33.5$ ft³
22. $V \approx 5086.8$ ft³ 23. 9.2 cm (rounded) 24. $y = 12$ cm; $z = 6$ cm 25. Linear units like cm are used to measure perimeter, radius, diameter, and circumference. Area is measured in square units like cm² (squares that measure 1 cm on each side). Volume is measured in cubic units like cm³.

Cumulative Review Exercises: Chapters 1–8 (page 613)

1. *Estimate:* $300 + 60,000 + 6000 = 66,300$; *Exact:* 64,574
2. *Estimate:* $20 - 10 = 10$; *Exact:* 10.242 3. *Estimate:* $4 \times 7 = 28$; *Exact:* 26.7472 4. *Estimate:* $30,000 \times 70 = 2,100,000$; *Exact:* 2,074,587 5. *Estimate:* $600 \div 3 = 200$; *Exact:* 200.8
6. *Estimate:* $5000 \div 50 = 100$; *Exact:* 94
7. *Estimate:* $5 + 5 = 10$; *Exact:* $9\frac{2}{5}$ 8. *Estimate:* $3 - 2 = 1$; *Exact:* $1\frac{7}{24}$ 9. *Estimate:* $3 \cdot 2 = 6$; *Exact:* $5\frac{1}{3}$ 10. $\frac{9}{20}$
11. 0.9132 12. 70 R79 13. 32 14. 0.000078 15. 39,105
16. 19.44 (rounded) 17. $4\frac{5}{12}$ 18. 836.085 19. 9 20. 38
21. two hundred eight ten-thousandths 22. 660.05 23. 2.055; 2.5005; 2.505; 2.55 24. *Per* means divide and *cent* means 100, so divide by 100 to change a percent to a decimal. 25. $\frac{1}{50}$
26. 2% 27. 1.75 28. 175% 29. $\frac{2}{5}$ 30. 0.4 31. $\frac{8}{1}$ 32. $\frac{7}{3}$
33. 35 34. 100 35. 2.01 (rounded) 36. 160% 37. $600
38. 135 min 39. $2\frac{1}{2}$ or 2.5 lb 40. 0.08 m 41. 1800 mL
42. mm 43. L 44. kg 45. cm 46. 0 °C and 100 °C
47. $P = 11$ in.; $A = 7$ in.² 48. $P = 5.8$ m; $A = 1.575$ m²
49. $C \approx 31.4$ ft; $A \approx 78.5$ ft² 50. $P = 76$ cm; $A = 264$ cm²
51. $P = 40.8$ m; $A = 95$ m² 52. $P = 50$ yd; $A = 142$ yd²
53. $y \approx 24.2$ mm 54. $x \approx 8.1$ ft; $P \approx 19.1$ ft
55. 59% (rounded) 56. Brand T at 15.5 oz for $2.99 − $0.30 coupon 57. $V \approx 2255.3$ cm³

58. $P = 52$ in.

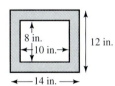

59. $1\frac{1}{12}$ yd 60. 42.9 lb (rounded) 61. 3.4 m 62. $22.40
63. (a) $0.55; $0.55 (b) $1.43; $1.39 (c) $2.53; $2.44
64. (a) 3.0% increase (b) 4.5% decrease (c) 0.0% (no change)
(d) 23.4% increase 65. No; 36 oz is more than 2 lbs, and the $3.95 priority rate applies to a package weighing 2 lbs or less.
66. $4442.13